Applying Local Climate Effects to Homicide Investigations

Applying Local Climate Effects to Homicide Investigation presents the concepts behind using local climate and weather records to enhance understanding of criminal cases. While sources of such local climate and weather information vary by country and regions, weather conditions are typically measured at airports or grassy areas as part of national, regional, or state-wide networks using many different instruments.

The information derived from such instruments and weather reporting services and agencies can inform and impact investigations, especially in the cases of natural death or homicide. The determination of post-mortem interval (PMI) in homicide cases is often based on entomological or anthropological evidence in combination with local climate estimations. Determining the local climate conditions typically requires knowledge of the environmental conditions where the body is found and the conditions where the measurement record was made. Most people recognize that cities are hotter than the surrounding countryside and that lake and coastal shorelines are cooler than fields: all of these comprise local climates. The local climate where a corpse is discovered usually differs from where temperature and humidity measurements are made. Consequently, many investigators and forensic pathologists do not realize the influence that such local temperatures and humidity can have on post-mortem interval determinations.

The book focuses on local climate conditions associated with the determination of post-mortem interval (PMI) and gives concepts behind adjusting climate information for local climates at the corpse. In addition, the book presents guidelines for crime scene investigators and lawyers to determine whether or not expert consultation is needed, and whether or not on-site measurements are needed. Most importantly, this book presents tools (data sources and modeling approaches) to guide forensic climatologists conducting forensic climatology work. It offers a basic, working understanding of the influence of the local environment on the local climate for forensic entomologists, forensic anthropologists, crime scene investigators, and lawyers. Numerous case studies are included throughout to show approaches, illustration decision points, and provide an understanding of the various impacts of regional and micro-climates upon decedents and their remains.

Applying Local Climate Effects to Homicide Investigations

Richard H. Grant

CRC Press is an imprint of the
Taylor & Francis Group, an **informa** business

Designed cover image: Shutterstock

First edition published 2025
by CRC Press
2385 NW Executive Center Drive, Suite 320, Boca Raton FL 33431

and by CRC Press
4 Park Square, Milton Park, Abingdon, Oxon, OX14 4RN

CRC Press is an imprint of Taylor & Francis Group, LLC

ISBN: 978-1-032-78187-7 (hbk)
ISBN: 978-1-032-78185-3 (pbk)
ISBN: 978-1-003-48663-3 (ebk)

DOI: 10.4324/9781003486633

Typeset in Minion
by SPi Technologies India Pvt Ltd (Straive)

Contents

Author

Richard H. Grant is a professor of agronomy and agro-micrometeorology at Purdue University who has consulted over the past 30 years in forensic climatology for 10 homicide cases in Ontario, Canada, Indiana, New Mexico, Illinois, North Carolina, Tennessee, Texas, Kansas, New Mexico, and Montana—in addition to having conducted climatological investigations for several non-homicide court cases, insurance claims, and environmental compliance issues. He has taught over 60 course-semesters on topics including climate, local climate, microclimate, forensic climatology, and meteorological science at Purdue University, State University of New York College of Environmental Science and Forestry and St. Joseph's College since 1983. He has authored over 170 articles and over 140 conference abstracts or proceedings addressing many aspects of local climates. While he has authored seven book chapters, this is his first published book.

Contributors

Leon G. Higley is a Professor of Applied Ecology at the University of Nebraska. He is an entomologist who has consulted on 26 cases including 20 homicides and has authored or edited five books, 22 book chapters and over 170 articles and conference abstracts.

Neal H. Haskell is Professor Emeritus of Forensic Science and Biology at St. Joseph's College. He is an entomologist who has consulted on over 1300 cases worldwide and has published one book, 20 book chapters and 86 articles and conference abstracts.

Preface

This book is not for experts. It is written for the crime investigators, prosecutors, and lawyers involved in homicide investigations and prosecutions. The motivation for this book came from a case I was consulting on several years ago. In that case, the Kansas crime scene investigators provided excellent supporting information for me but were surprised with my steady stream of information requests. So, they asked me 'Is this written up anywhere?'. My answer then was 'no'. This book makes that 'no' a 'yes.'

In this book, I seek to familiarize you with the science involved in determining the daily temperature and humidity where a body was found without 'burying' you in detail. You do not need to read the book sequentially, but it is helpful. I have added a list of abbreviations and a glossary of terms used that you may not be familiar with. The first use in each chapter of each abbreviation listed and word found in the glossary is highlighted with **bold** type. References are indicated with a superscripted number (like [10]) and listed at the end of each chapter.

Abbreviations

ADD	Accumulated Degree Days
ADH	Accumulated Degree Hours
AGL	above ground level
ASL	above average sea level
ASOS	Automated Surface Observation System
ASHRAE	American Society of Heating, Refrigeration and Air-conditioning Engineers
AWOS	Automated Weather Observation system
BTU	British Thermal unit
COOP	United States National Weather Service Cooperative 'summary of the day' weather station
CRN	Climate Reference Network
ECAD	European Climate Assessment and Dataset
F	Fahrenheit
HR	hour
MADIS	Meteorological Assimilation Data Ingest System
MAE	Mean Absolute Error
MBE	Mean Bias Error
MMTS	Maximum Minimum Temperature system
NIST	National Institutes of Standards and Technology
NOAA	United States National Oceanic and Atmospheric Administration
NCDC	United States National Climate Data center
NCEI	National Centers for Environmental Information
NWS	United States National Weather service
PMI	Post-Mortem interval
R	Correlation coefficient
R2	Coefficient of determination
RMSE	Root Mean Squared Error
RWIS	Roadway Weather Information systems
WMO	World Meteorological Organization

Introduction and overview 1

When did the homicide victim die? The changes in the condition of a corpse over time depends largely by the physical and biological conditions right where the corpse lies. So, estimating the post-mortem interval (**PMI**) based on the insect development (Chapter 2) or decomposition state (Chapter 3) should consider the **climate** where the corpse is found. Even though the PMI does not usually play a dominant role in court,[1] it is useful in assessing suspect's alibis.

But how does that climate influence a body? Let's start out considering you or I sitting or lying somewhere. Our skin **temperature** is an equilibrium between heat coming from the body core and heat coming from or going to the environment (discussed in Chapter 4). Our bodies also loose water to the environment by breathing and sweating, limited by the **humidity** of the environment (discussed in Chapter 5).

On death, the body begins to decay and the core temperature cool. This cooling typically takes 24 to 48 hours.[2, 3] At this point in time, the skin temperature becomes controlled by the environment.[4, 5] Insects may be attracted to the corpse within minutes of death with subsequent oviposited eggs hatching within a day or two.[6] Egg hatching and Insect development can be estimated from the heat available for metabolic activity. This available heat is expressed as Accumulated Degree Hours (**ADH**) or Accumulated Degree Days (**ADD**), accumulated from the time of oviposition to corpse discovery. Very warm climates promote rapid insect development while **cold climate**s inhibit insect development. The decay of a corpse is strongly influenced by the moisture and temperature of the corpse surface and as a result the humidity and temperature of the air.[6] While exhaled breaths and sweating cease on death, water continues to be lost from the corpse through open wounds and body orifices in relationship to the body and air temperature and air humidity. Very moist warm climates promote corpse decay while very dry environments inhibit corpse decomposition and promote desiccation and mummification.

Any assessment of a corpse climate is an *estimate*- bounded by an **error** of the estimate. As such you will find within descriptions of the **accuracy** and errors typically occurring from the use of measurements, and models to estimate the corpse climate time series (Chapters 7, 9, and 10). The common overestimates of PMI in warm climates and underestimates in cold climates[7]

DOI: 10.4324/9781003486633-1

are at least partly due to a lack of understanding of how to apply available climate information to the climate around where the body was found. The air temperature and humidity around a corpse depends on heat and water exchange with its immediate surrounding environment, called the '**microclimate**'. The microclimate is influenced by terrain influences on the day-to-day climate, called the '**local climate**' (discussed in Chapters 4 and 5). These local climates are then influenced by larger-scale influences of **weather** systems as commonly reported in the news and seen in weather maps. An example of how these different scales of climate interact is the microclimate under a forest canopy in vicinity of a reservoir (a local climate) as affected by warm moist southerly winds ahead of a cold **front** passage.

So, in what kinds of microclimates are most corpses found? A study of North American homicides found about one-third of corpses were discovered in buildings, a third in open public areas and a third in bushes, ditches, trails, under bridges, beside side roads or in alleyways or shallow graves.[7] In a rural Finland survey, roughly three-quarters of the corpses are found in woodlands and one-quarter in water.[8] In Japan, about a third of corpses from mutilation homicides are found in the forest, a third in water bodies, and a tenth in urban areas.[9] Corpses are often concealed from the casual observer: surveys have shown that 43% of corpses were concealed by snow, plants or burial in rural Finland[8] while only 15% of homicides were concealed by bushes, buried, or dumped in a river in India.[10] The characteristics of these and other microclimates are discussed in Chapter 6.

Case studies describing estimates of ADD or ADH in a range of microclimates and local climates are presented in the Appendix. ADD or ADH are usually estimated from modeling the microclimate energy budget (Chapter 6) which adjusts the **record**ed conditions at a nearby climate station (Chapter 7). Sometimes additional post-discovery on-site measurements are needed to make these models (see Chapter 8). The sources and accuracy of climate information and modeling of microclimates used in estimating ADH or ADD as well as that of instruments used to measure post-discovery on-site air temperature and moisture and are discussed in Chapters 7, 8, and 9.

References

1. Madea, B. and Henssge, C. 2023. pp.1–7 In: Madea, B. (Ed), *Estimation of time since death*. 4th Ed., CRC Press, Boca Rotan, FL, USA.
2. Muggenthaler, H., Sinicina, I., Hubig, M. and Mall, G., 2012. *International Journal of Legal Medicine, 126*, pp.79–87.
3. Mohiddin, S.K., Hussain, A.J., Kumar, V., Subhedar, A. and Khan, M.T., 2019. *Journal of Indian Academy of Forensic Medicine, 41*(1), pp.34–37.
4. Smart, J.L. and Kaliszan, M., 2012. *Legal Medicine, 14*(2), pp.55–62.

5. Abraham, J., Wei, T. and Cheng, L., 2023. *Journal of Forensic Sciences*, *68*, pp.884–897.
6. Janaway, R.C., Percival, S.L. and Wilson, A.S., 2009. pp.313–334 In: Perceval, S.L. (Ed.) *Microbiology and aging: clinical manifestations*. Springer, LLC.
7. Chapman, B., Raymer, C. and Keatley, D.A., 2022. *Homicide Studies*, *26*(2), pp.199–215.
8. Flores, V., Kim, H., Sielawa, M., Malinowska, P., Ramanauskas, B., Becker, D., LeRoux, H., Häkkänen, H., Hurme, K. and Liukkonen, M., 2007. *Journal of Investigative Psychology and Offender Profiling*, *4*(3), pp.181–197.
9. Zaitsu, W., 2022. *Journal of Forensic Sciences*, *67*(6), pp.2367–2375.
10. Mohanty, M.K., Mohanty, S. and Acharya, S., 2004. *Medicine, Science and the Law*, *44*(2), pp.160–164.

How local climate affects PMI estimates using entomological methods

2

NEAL H. HASKELL
AND LEON G. HIGLEY

The use of forensic entomology to assist in both civil and criminal courts has been available for well over a century. These applications ranged from resolving disputes over time and proper use of insecticide applications to determining when a person has died. This later use, estimating the time of death, has increased dramatically over the past 40 years requiring hundreds of novel research projects being published. What this research has identified are three major variables which must be known to the forensic entomologist to enable him to draw reliable conclusions for his case.

These three variables include the kind of insect (the species name or at least the closely related insect taxa having very similar growth and development); the oldest life stage of the identified kind of insect; and the **temperatures** (temperature **data** or a reasonable estimation of the temperatures) present from where the remains were recovered or locations where the remains were suspected to have been. This last variable, temperature, *must be* documented from either **weather (climate)** station data or estimated using knowledge of the scene environment, time of year, and geographic location. Without this critical value, the time of death *cannot be* determined! *Temperature is the key.*

Forensically, insects are both bane, through destroying evidence, and boon, through being evidence themselves. Most commonly, insects associated with dead bodies are useful in estimating the **postmortem interval (PMI)** – the time since death. Because flies rapidly find dead bodies under most circumstances and lay eggs on the body, the age of maggots (the developing flies) found on a body have the potential to indicate the time of egg laying and by association the time of death. But to reiterate from above, there are three crucial factors: (1) the species of insects, (2) the stage of the insect, and (3) the temperatures the insect has experienced.

Insect development depends on time and temperature, with higher temperatures resulting in faster growth. The relationship between growth and temperature vary with species and by stage, hence their importance. Temperature is the driving force for insect growth. So, the more accurate our

DOI: 10.4324/9781003486633-2

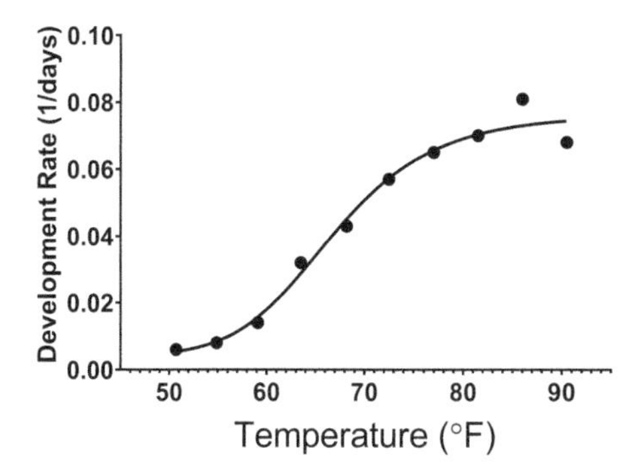

Figure 2.1 Heat required for completion of *L. sericata* development from the egg stage. A typical growth rate vs. temperature curve, here for the egg stage of the blow fly *Lucilia sericata*. The solid line is a spline curve fit of the measurements for illustration purposes only. Data and figure provided by N.H. Haskell and L.G. Higley.

estimate of the temperatures the growing insects experience, the more reliable our estimate of the PMI. Generally, this relationship between growth rate and temperature approximates an s-curve, as illustrated in Figure 2.1.

Obviously, there are limits to the role of temperature and growth. If temperatures are too cool, as in winter, late fall, or early spring, insects can't grow. If temperatures are too hot, the rate of insect growth may be limited or even decrease. Other factors can limit growth such as sun and rain, as well as **relative humidity**. The key issue with **humidity** is if conditions are so dry that the dead body starts to dry or even mummify, it becomes unsuitable for fly egg laying and development. Finally, rain and sun may force maggots to move to sheltered locations on the body, which also may enhance or delay development.

Besides weather, other factors can delay egg laying and must be considered in estimating a PMI. First, certain species of blowflies usually have a more than a 12-hour delay in egg laying (another example of why knowing the species of blowfly is crucial). Second, barriers to flies accessing a body can also result in delays. Such barriers include wrappings, bags, and bodies inside cars or houses. None of these barriers are likely to prevent egg laying eventually, but they can delay it by hours or even a day. Third, daily temperatures matter. Below a certain threshold temperature, which varies by species, blowflies are unable to fly. Thus, if daily temperatures are too cold, flies are unable to reach a body and lay eggs, this egg laying is delayed until temperatures warm up. It is also important to note that blowflies don't lay eggs at night, so it is daily (diurnal) temperatures that matter for egg laying.

In a very unusual forensic entomology case from Wisconsin, a women's body was found on January 22, lying face down on a gradually sloping hill side. There were tall grass and scattered weeds surrounding the remains. The body showed very little decomposition and was definitely recognizable from her facial appearance. She was clothed in workout sweats from the waist downward, with a long sleeve sweater covering her upper torso. She had on socks, but no shoes. The socks were not soiled, as if she been dumped at the location and not walked in her stocking feet. The next day at autopsy, during the external examination of the remains, the coroner searched the remains thoroughly for forensically important insect life stages. Finding none he concluded that the period she had been out was a very short period of time. The only insect specimens recovered were two adult insects found when examining her hair. These two insects were a field ant and a ground beetle; neither of the would have been forensically involved. They are just two common ground crawling insects that had to be somewhere. When this entomological evidence was received by the forensic entomologist, he stated, "What can I do with these two specimens"? That was before he consulted the weather data. Through the winter months in Wisconsin temperatures are cold! He began working backwards day by day studying the maximum and minimum temperatures, ultimately discovering that there was not one day during the month of January or December when temperatures were above freezing. It wasn't until the 25th of November that there was one day when temperatures reached high enough to allow insect activity. Thus, an ant and a ground beetle decided to take a walk through the victims 's hair and then became trapped in her hair when temperatures went below freezing. She was eventually identified. Her husband stated to police that he had become frightened when she died from alcohol and drugs and decided to dump her body. He forgot to tell the police that he had wrapped his arms around her throat and strangled her. So in the end, the daily temperatures from Wisconsin proved when these two insects could only have been walking about for *only one day*. This is the very reason that temperatures are so very critical to the forensic entomologist's analysis.

In principle, the location on the body of developing maggots will influence the temperatures they experience. In practice, the small variation in temperatures on the body is much less than the variation in temperatures used to estimate development (see Chapter 6). Here, the principle applies that small **error**s don't matter until large errors are removed. This large error is the difference between temperatures at the body versus temperatures at the nearest temperature **record**ing site.

Accounting for this difference is important to avoid introducing **bias** in our estimates of insect development. Various approaches to estimating temperature are discussed in Chapter 9. A further consideration is the weather during the interval while insects were developing. If a weather **front** moves through during this period, relationships between scene and recording site temperatures may change. In this instance multiple estimates of the scene/recording station temperatures are necessary to reflect the changes associated with different frontal systems (see Chapter 4). Failing to account for such differences between scene and recording station temperatures can be substantial.

A case study will illustrate this statement. The case from Sacramento, California during the late 1980s involved a 10-year-old boy who was murdered with the body being dumped in brush on the banks of the Sacramento River. Temperature data for the period of time the boy was missing were recovered from the National Weather Service Automated Surface Observation System (**ASOS**) station, Sacramento Airport which was approximately 20 miles from the scene. Once the insect evidence was examined there were major time discrepancies between the time he had last been seen and the age of the maggots. It was then that the forensic entomologist suggested to the crime scene technicians to return to the scene and record hourly temperatures five or six times per day (discussed in Chapter 8). They were also directed to obtain the same hourly temperatures from the ASOS station at the airport (see Chapter 7). These temperatures were collected from both the death scene and the ASOS station for about a week. The differences in temperatures were dramatic with afternoon temperatures at the ASOS station exceeding 20 degrees **F**. Night time temperatures were somewhat similar between sites. With this adjustment a correction factor of between two and three weeks was added to the PMI.

The correction factor varies with the temperatures. With warm temperatures and rapid insect development, the difference may be a day or two. But with cool temperatures and slower insect development, differences can be a week or weeks. Most generally, as illustrated in Figure 2.2, if the ASOS station is close to the crime scene, the correction in observed temperatures may only be less than 3° F (1.6° C). In this example, the location of the body was less than five miles from the ASOS station, and no frontal systems moved through

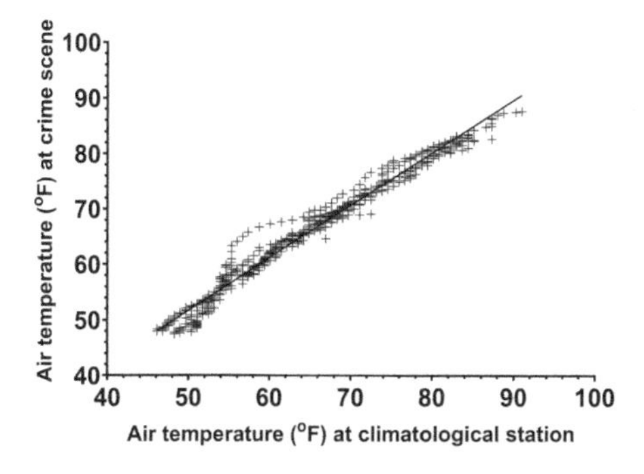

Figure 2.2 Comparison of air temperatures measured at a climatological station to those measured on-site. The solid line indicates the linear correlation between the hourly on-site temperatures (as measured on-site post-discovery) from the ASOS station temperatures.

Data sources: NOAA National Center for Environmental Information, Automated Surface/Weather Observing Systems (ASOS/AWOS) and Concordia University, Seward, Nebraska. Figure provided by N.H. Haskell and L.G. Higley.

the area during the time the body was exposed. Consequently, the calculated PMI from corrected temperatures only differed by a few hours from that calculated with uncorrected values.

Blow flies go through multiple stages during their development: egg, three larval stages, pupa, and adult. The egg, first larval stage, and second larval stage are all of relatively short duration. In contrast, the third larval stage represents the longest of the developmental stage. During the first half of the third stage maggots complete approximately 90% of the total larval feeding. Once feeding is completed, larvae of most species enter a wandering, or migratory, phase. During larval migration maggots may move up to 30 feet away from a body or more, with migration under some circumstances having been measured up to ca. 350 ft. Eventually, migrating maggots stop and enter the pupal stage after moving into the soil or under vegetative **litter**. Unlike other aspects of larval development, the migration period is only time dependent and does not depend on temperature.

Another aspect of the third larval stage is that blowfly maggots feed in very large groups or masses. While in a mass, maggots generate so much metabolic heat that mass temperatures become substantially higher than ambient. Indeed, in many masses the temperatures are sufficiently high to become lethal. Individual maggots avoid death by moving from the middle to the outside of a maggot mass in an effort to cool. How does this movement, during which maggots stop feeding, influence development times? And how should we reflect the maggot mass temperature when we are estimating development? Neither of these questions have a clear answer. It is obvious that using the mass temperature can lead to overestimating the rate of larval development. On the other hand, once maggots are in a mass their temperature environment is independent of ambient temperatures. However, it is only when they are in the very late period of the 2nd stage and during the feeding stage of the 3rd stage when this constant temperature-controlled environment exists. In determining the PMI most forensic entomologists, once larvae are in a mass, estimate their thermal environment by a single constant temperature, typically the temperature at which the growth rate is maximum.

How is the PMI determined based on species, stage, and temperatures? The process is essentially a trip back in time, using known information to backtrack and determine when it was most likely insect eggs were laid. The starting point is the stage of the insect found on a body. For example, at −77 °F the eggs of most insect species required about 12 hours to hatch. Thus, if a body at room temperature has unhatched eggs on it, a good starting estimate is that the PMI is less than 12 hours. If the stage on the body is larval (and most often it is the third larval stage), our back tracking process is more complex. Each species of forensically important fly has a unique relationship between growth and temperature. Through research, these relationships are mathematically modeled (usually by linear regression). These models relate temperatures to growth rate, and they let us calculate degree days. Degree days are a combination of temperature and time; in their simplest form they represent the number of degrees above some minimum for a period of one day. Each stage of an insect has a different number of degree days (which is a product of both temperature and time; not just temperature, not just time) required for growth to the next stage. So, by using our estimates of the temperatures the larvae experienced we can calculate degree days and estimate when each stage was completed, all the way back to the egg stage and egg laying. This discussion slightly simplifies the actual process. In practice each stage has a different degree day requirement, a different threshold temperature for calculating degree days, and different factors that may influence temperatures, like maggot masses, for example.

Can we estimate a PMI without complete information on species, stage, and temperature? It depends. Sometimes the only evidence available from a homicide are pictures of insects on a body. From these we might be able to estimate the stage, and then make conservative estimates of development times if temperature information exists. And because we may not know the species involved, we may still be able to offer a PMI estimate by using data from a species with the fastest growth rates, thus providing a PMI estimate that is of at least a minimum duration, but possibly longer. And what about temperature data? If data are incomplete, we may be able to make informed estimates of temperature ranges and still offer a crude estimate of the PMI. However, if all temperature information is missing, a PMI estimate from insect development becomes impossible.

How local climate affects PMI estimates using anthropological methods?

3

Forensic anthropology seeks to identify a corpse or skeletal remains, determine the biological characteristics of a corpse, and estimate the **PMI**. We focus here on determining the PMI when the corpse core **temperature** varies with the environmental temperature- typically beginning at least 24 to 48 hours after death. Determining the PMI after this point in time involves understanding what influences corpse decomposition. While the stages of corpse decomposition are generally the same,[1] the time it takes to reach any given stage is highly variable. The duration of each stage, the rate of decomposition, depends greatly on the environment where the body is disposed-above-ground, below-ground, or in water.

Temperature and moisture strongly influence corpse decomposition.[2] And since air temperature and moisture depend in part on the influence of the large-scale **weather**, the **local climate**, and the actual **microclimate** where the body is found, these environmental conditions need to be considered in PMI estimation.

Temperature is usually recognized as the dominant control on decomposition rates because the available heat influences the microbial activity in the corpse. We can use the influence of temperature on decomposition to estimate PMI. One approach is to determine the heat needed for the observed decomposition (represented by the accumulated degree days, or ADD) and then back calculate the **ADD** from the day of discovery. Various schemes have been developed to estimate the ADD needed to result in observed corpse decomposition: the total body score (TBS),[3,4] total body score surface (TBS_{surf}),[4, 5] and total decomposition score (TDS).[6] Assuming these estimates of required ADD for the amount of observed decomposition are correct, the PMI is determined back-tracking in time the ADD at the corpse based on estimated corpse microclimate.

The first problem in this approach is knowing the corpse microclimate. Direct use of nearby climate **data** does not generally represent the conditions at the corpse.[7] A second problem is that these models assume all possible corpse conditions, corpse environments are represented in the model. Marhoff-Beard and coworkers[8] found that exposure of the corpse to sunlight

 DOI: 10.4324/9781003486633-3

influenced the **accuracy** of the TBS model- and clearly the corpse may be only sunlit part of the day depending on the local environment (trees, hills, bushes, etc). The TBS model overestimated the PMI in temperate Australian climates.[8]

But decomposition is also influenced by **humidity**- or rather a moist environment for the microbial populations on and in the corpse. In a survey of corpse decomposition over time, Myburgh and coworkers[9] found that decomposition occurred much faster in the summer than winter- even when the ADD were the same. They thought this was because the air was drier during the winter than summer, with dry conditions inhibiting microbial activity.

Can a better job of estimating PMI come from including both a measure of temperature and moisture at the corpse location? Such a PMI estimation model was developed for above-ground or below-ground corpses by Vass.[10] This model has also been shown to have **error**s related to climates. Surveys of deaths with known PMI found the estimated PMI using this 'universal' model was often over-estimated in warm climates, underestimated in **cold climate**s, and underestimated in **temperate climate**s.[8, 11] Part of this estimation error is likely due to the use of temperature and humidity measures of either that of the day of discovery or that averaged over a period of time as stipulated by Vass.[10] For a day to represent the entire PMI period, one has to assume the weather does not change over the PMI period- clearly not generally true. For a nearby climate station to represent the conditions at the corpse, the two locations need to be nearly identical- also generally not true.

So in summary, while the qualitative estimates of decomposition used in the PMI models continues to be a challenge that the anthropologists need to address, a better representation of the temperature and moisture environment where the corpse is found is also needed. Significant errors in PMI based on the decomposition will continue to exist when nearby climate data are used without considering the differences between where the corpse is found and where the climate measurements are made. *One needs to know the very local temperature and moisture conditions at the corpse, not just the conditions at some nearby climate/weather station, to assess PMI using observed corpse decomposition.*

References

1. Goff, M.L., 2010. pp.1–24 In: Amendt, J., Goff, M.L., Campobasso, C.P., Grassberger, M., Eds., *Current concepts in forensic entomology*, Springer.
2. Janaway, R.C., Percival, S.L. and Wilson, A.S., 2009. pp.313–334 In: Perceval, S.L. (Ed.) *Microbiology and aging: clinical manifestations*, Springer LLC.
3. Megyesi, M.S., Nawrocki, S.P. and Haskell, N.H., 2005. *Journal of Forensic Sciences*, *50*(3), p.JFS2004017.
4. Smith, D.H., Ehrett, C., Weisensee, K., and Cristina Tica, C., 2016. *Journal of Forensic Sciences*, *61*(Suppl 1), pp.S201–S207; *Journal of Forensic Sciences*, *68*(1) (2023): pp.355–358.

5. Moffatt, C., Simmons, T. and Lynch-Aird, J., 2016. *Journal of Forensic Sciences, 61*, pp.S201–S207.
6. Gelderman, H.T., Boer, L., Naujocks, T., IJzermans, A.C.M. and Duijst, W.L.J.M., 2018. *International Journal of Legal Medicine, 132*, pp.863–873.
7. Dabbs, G.R., 2010. *Forensic Science International, 202*(1–3), pp.e49–e52.
8. Marhoff-Beard, S.J., Forbes, S.L. and Green, H., 2018. *Forensic Science International, 291*, pp.158–166.
9. Myburgh, J., L'Abbé, E.N., Steyn, M. and Becker, P.J., 2013. *Forensic Science International, 229*(1–3), pp.165–e1.
10. Vass, A.A., 2011. *Forensic Science International, 204*(1–3), pp.34–40.
11. Cockle, D.L. and Bell, L.S., 2015. *Forensic Science International, 253*, pp.136–e1.

Controls on local climate temperature

4

The **temperature** of our **local climate** influences our body temperature in the environment. When we think about being comfortable, we usually mean that we do not feel too hot or too cold. The human body has a heat source keeping our 'core' temperatures around 98.6 °F. When we have a fever, our bodies' increase the production of heat causing us to be hotter than normal. So, heat being produced in our core affects our temperature. If we are cold, we usually put on more clothing, when too warm we usually take off clothing. The clothing reduces the heat movement from or towards our bodies- providing **insulation** for our body to minimize heat moving to or from the environment. The movement, or transfer, of heat between our body and the local climate of our environment affects our body temperature. The amount of heat something has is measured by temperature.

The relationship between the temperature *of* a body and the heat contained *in* the body is described by the **heat capacity** of the material that the body is made of. The units of this are in Heat per unit volume (heat capacity). The greater the heat capacity of something, the more heat the object can hold for a given temperature. The heat capacity of a material describes the heat contained in a cubic foot of material (**BTU** ft^3). Because air, water, and concrete have different **heat content**s the heat contained in a cubic foot of the material will differ even when they are at the same temperature: if at 50 °**F**, the air contains about 1 BTU ft^3, the water contains 3120 BTU ft^3 while the concrete contains 1375 BTU ft^3. There is much more heat stored in water and concrete or air at the same temperature, resulting in an influence of water bodies and urban areas on local climates. The variations in heating and cooling over hours or days then can be expected to have different effects on the air, water, ground and building temperatures in inverse relation to their heat capacities. These differences result in local climates discussed below.

The male and female human body are respectively about 51% and 44% water by weight.[1] Besides water, the primary materials in humans include bone, muscle, fat, and skin. The heat capacity of the largest component of the body, the abdomen is approximately 88% that of water.[2] Exchange of heat between a corpse and the surrounding environment can be described by an **energy balance** at the body surface, with heat moving between the core of the body and the surface and heat moving between the body surface and the local environment.

DOI: 10.4324/9781003486633-4

The energy balance

The temperature of a body represents the **'sensible' heat content** of the body, or the heat that can be sensed or measured with a **sensor**. There is energy contained in a body that cannot be sensed called **'latent heat'**. Latent heat is energy that is bound in the molecular structure of water. The combination of sensible and latent heat of water describes the total energy of the water in whatever state it is in- solid, liquid, or gas.

The local environment can be thought of as a distribution of materials with different properties that exchange heat resulting in a temperature of each material. The temperature of a surface is the net effect of the transfers of both sensible and latent heat to and from the surface. When the energy transfers coming to a surface are equal to those going out, the surface is in equilibrium and the air temperature and surface temperature is steady. This is called an energy balance. When this is true, there is a steady body temperature and steady surface temperature. Simply put:

$$\begin{aligned}\text{total energy in the body} = &\ \text{energy stored in the body}\\ &+ \text{energy into the body}\\ &- \text{energy out from the body}\end{aligned}$$

For a living human, metabolism seeks to maintain the heat stored in the body at a steady amount resulting a body core temperature of approximately 98.6 °F. If the living body temperature is rising due to the influences of the environment (more heat into the body than out from the body), it adapts to increase heat loss by decreasing metabolic activity, increasing the transfer of heat from the body core to the surface including vasodilation (increasing blood vessel diameter) and panting, and evaporating water at the skin (sweating). Likewise, if the body temperature is cooling due to the environment, it adapts to increase metabolic heat production and reduce heat loss from the body through vasoconstriction (shrinking blood vessel diameter). So, we see that the actual rate of cooling is dependent on the state of the body and the local environmental conditions.

A corpse however has negligible metabolic heat and no blood flow, so it will warm or cool (depending on the temperature **gradient**) until it reaches the same temperature as the environment (has reached an energy balance with the environment). While the previously healthy human body initially starts at a core temperature of around 98.6 °F and skin temperature of around 92 °F, the metabolic heat sources in the body slowly cease and heat is transferred from the core to the skin of the corpse resulting in the initial body cooling. The high amount of stored heat in the body and the relatively slow transfer of heat from the core to the skin results in a gradual cooling of the body.[3] This initial cooling of a corpse to a steady temperature of the environment in **still air** increases with the corpse mass to around two days.[4] The rate of cooling however is also influenced by the environmental temperatures.[3, 5]

The environment of a corpse also has stored heat (measured by its temperature). The transfers of **sensible heat** to and from a body to and from the local surroundings include that by **conduction**, **convection**, and **advection**. Heat is also transferred as radiant heat to the body from the sun (called **solar radiation**) and sky (called **sky radiation**). Heat is also transferred to and from a body as radiant heat due in part to the temperature of the environment and the body- termed here '**thermal radiation**'.

The energy exchanges for a body (illustrated in Figure 4.1) can then be restated as:

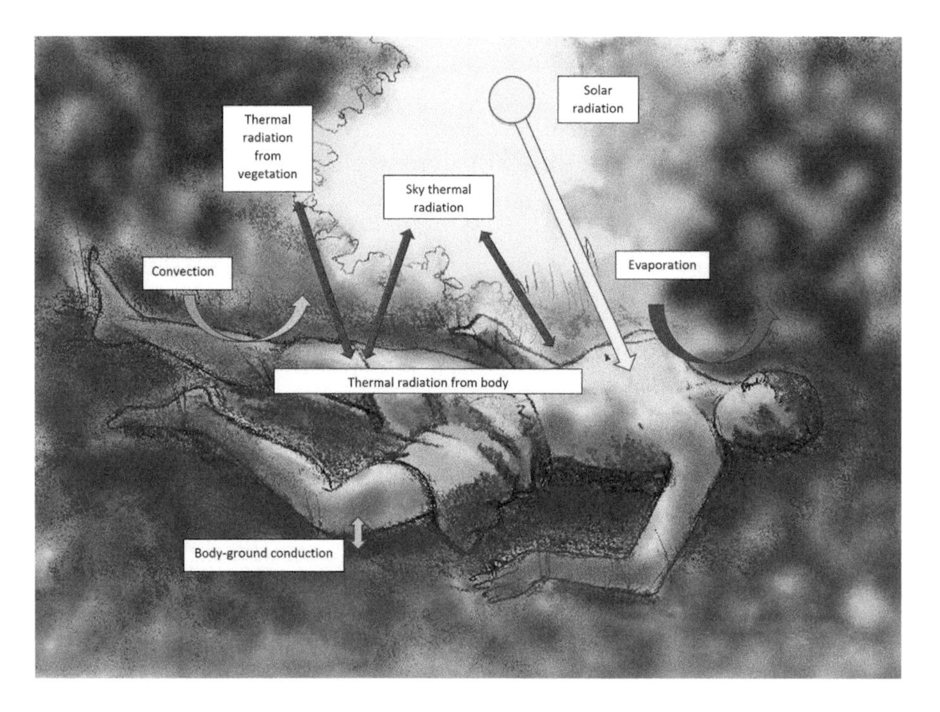

Figure 4.1 Energy exchanges between the air and ground and the corpse. Energy exchanges of thermal radiation (red arrows), Solar radiation (yellow arrow), conduction (green arrow), convection (orange arrow), evaporation (blue arrow). Illustration by E. F. Grant.

$$
\begin{aligned}
\text{total energy in the body} = {} & \text{energy stored in the body} \\
& + \text{radiation exchange energy with the body} \\
& \quad (\text{thermal, sky and solar radiant heat exchange}) \\
& + \text{convection exchange energy with the body} \\
& \quad \left(\begin{array}{l}\text{sensible heat moving with the movement} \\ \text{of the air}\end{array}\right) \\
& + \text{conduction exchange energy with the body} \\
& \quad (\text{sensible heat moving through solids})
\end{aligned}
$$

Since both the body and air contains water, there is an additional means of sensible heat transport resulting from the **evaporation** of liquid water or the **sublimation** of solid water (ice) (Figure 4.1). Chapter 5 discusses the sensible heat transports associated with changes in the state of water (solid, liquid, or gaseous).

Daily variation of air and water temperatures

We generally assume that the body temperature and air temperature are constant over a given hour. The energy balance of a corpse with the environment results in an hourly average surface temperature and an hourly average air temperature. The air temperature just above the surface is often assumed to be equal to that at the surface. This will be discussed below.

Typical daily variation in air temperature (a diurnal pattern) is cyclical with higher temperatures during the day and lower during the night. Typically, as the sun nears the horizon the air temperatures decrease and continue to decrease throughout the night with the lowest temperatures just before sunrise. Maximum air temperatures typically occur in the early afternoon after the sun is the highest in the sky (Figure 4.2).

When we estimate the average daily air temperature, it is typically done assuming a symmetrical daily air temperature 'wave' pattern so the minimum daily air temperature and the maximum air temperature is separated by 12 hours (Figure 4.2). If the air temperature variation was truly a wave, the average of the maximum and minimum temperature would equal the average daily temperature. The average daily temperature was 1% lower than estimated using the maximum and minimum temperatures.

Since water has a much higher heat capacity than land, it will heat up and cool down slower than the land. The daily range in lake surface water temperatures is influenced by area. An assessment of 100 lakes found that lakes with areas of 2 acres had ranges of daily near-surface temperatures around 7 °F while lakes 250 acres or more varied in temperature by less than 2 °F

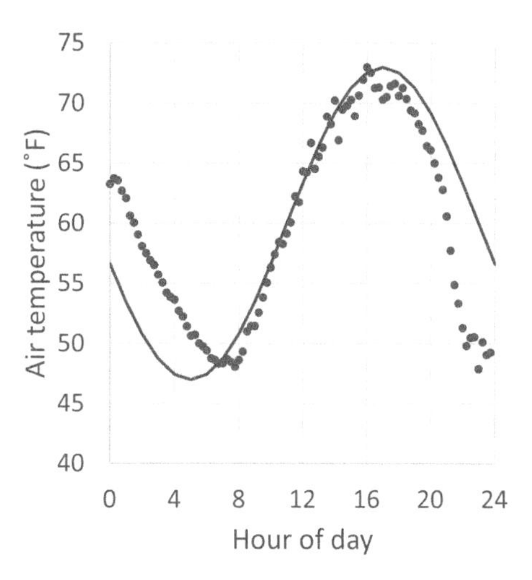

Figure 4.2 Typical daily variation in air temperature on a clear sky day in Indiana during May. The actual (blue circle) and modeled wave (red line) temperatures are indicated.

during the summer.[6] The water temperature in streams and rivers also follows a daily sinusoidal wave pattern like that of the air. However, since the water has a higher heat capacity than the air, the maximum and minimum daily temperatures are delayed, or lagged, from the time of the corresponding maximum and minimum air temperature.[7]

Radiative heat exchange

Radiated heat is most of what you feel when you are in front of a fire in the fireplace or a bonfire in the open. This is not the **radiation** associated with nuclear energy. The most important source of heat for the earth's surface or any object near the earth's surface is the heating by the sun. The sun heats bodies both directly by the beams of the sunlight (called direct beam radiation) but also by the light coming from reflections of sunlight by the gases and particles in the sky (called sky radiation) (Figure 4.1).

Solar radiation reaching the earth's surface varies with the **cloud cover**, atmosphere composition, and angle of the sun. Cloud cover reduces the annual incoming solar radiation by 26% over the whole earth.[8] Solar radiation reflects off clouds making them bright. Solar radiation reflecting off molecules and particles in the sky making us able to see the sky- and that reflected solar radiation is also received on the earth surface. Typically, the radiation from the sky and clouds reaching the ground is more than twice that received

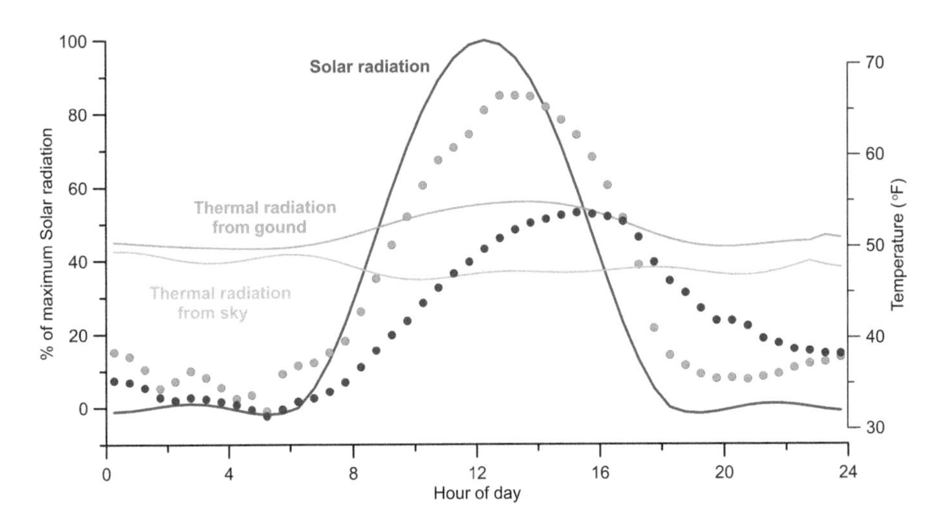

Figure 4.3 Influence of absorbed and emitted radiation on surface temperature of a grassland in Kansas on a clear sky day. Air temperature (blue circle) and ground surface temperature (green circle) are related to the solar radiation (red line), modeled thermal radiation from the sky(light blue line) and modeled thermal radiation from the ground (light green line).

Data source: Project supported by NASA under Contract NAS5-32356 to Alan Betts.

directly from the sun.[8] Partly-cloudy days do not reduce the solar and sky radiation received at the ground much: there is only about 5% more on clear sky days than partly cloudy days.

There is also heat radiated from everything in our environment. Every object that has a temperature radiates heat- hotter objects radiating more heat than cooler objects. So, a corpse on the ground receives radiated heat from the sun, sky, ground, trees and other vegetation, and buildings and any other natural or man-made surface feature and radiates heat relative its temperature- termed **thermal radiation** (Figure 4.3). On average the earth is warmer than the sky and as a result, the thermal radiation from the ground is 16% more than that from the sky. Cloud cover does not have a significant effect on the thermal radiation from the sky.

On a typical clear sky day, the solar and sky radiation reaching the ground surface is at a maximum near when the sun is highest in the sky while the ground is warmest in the afternoon (Figure 4.3). The thermal radiation **emitted** from the ground is highest when the surface temperature is highest (Figure 4.3). The combination of thermal radiation from the earth, earth objects, clouds and the air near the earth's temperature and solar radiation to or away from a surface is termed the **net radiation.**

Surface orientation effects on sunlight reception

The angle between the solar beam and the ground defines how much sunlight can be received on a surface. The closer the sun is to the horizon, the less sunlight is being received on any area of the ground. Solar noon is the time of day when the angle between the beam of the sun and the ground is the greatest for any given day. As the angle of the sun's rays to the ground increase the ground receives less **radiant energy** from the sun. In hilly terrain, the orientation of slopes can be important in influencing the amount of solar radiation reaching the ground. For east-west valleys or mountain ranges, the slopes facing south in the northern hemisphere will typically receive more sunlight than those facing north.[9, 10]

Buildings have surfaces at all angles relative to the ground: walls usually perpendicular facing the full range of compass directions and roofs commonly sloped. Since the solar radiation received depends in part on the angle that the sunlight beam is making with the receiving surface, the maximum heating of walls and roofs will differ from the ground. Walls facing east will receive the most sunlight in the morning and those in the west in the afternoon. North-facing walls in the northern hemisphere receive less sunlight than south facing walls. Since cloud cover commonly increases through the day, the surfaces receiving morning sunshine will likely heat more than those receiving afternoon sunlight.

For a corpse lying prone on the ground, the solar radiation is typically received by only 10% to 25% of the body depending on the posture of the body and the angle of the sunlight[11] (Figure 4.1).

Surface absorptivity effects

The amount of radiant heating of the ground or a building by the sun and the sky depends also on the ability of a surface to absorb the radiant energy. Only a fraction of the radiation reaching the surface of an object is absorbed by that object- contributing to the heat content of the object. This is called the **absorptivity** of the object—The sunlight that has been reflected, not absorbed, from surfaces. This is a surface's **reflectivity**. So dark roofs mean more sunlight (solar energy) is absorbed than light roofs. Increased surface absorptivity of sunlight on roof and pavement surfaces in Indianapolis, Indiana during Winter and summer show surface temperatures rise about 18 °F when the absorptivity increases 20%.[12]

Bare ground and vegetation absorb more radiant heat than snow covered ground. Snow absorbs much less solar radiation than bear ground.[13] Coniferous forest canopies absorb more radiant heat than deciduous canopies[14] resulting in a typically darker **understory** and less daily variation in temperature in the understory. Solar radiation penetrates shrublands and grasslands to a greater degree than forests,[14] allowing more sunlight to reach the ground.

The corpse also does not absorb all the sunlight reaching it. Bare skin absorbs 40% to 80% of the solar radiation reaching it, depending on skin pigmentation.[15] Similarly, clothing absorbs 38% (white clothing) to 76% (black clothing) of the received solar radiation.[16]

Sky view effects

Since the ground surface is not flat and typically has buildings and/or vegetation near where corpses are found, the amount of radiation from the sky and received by the corpse is not as much as that if the ground were flat and featureless. The proportion of sky that can be 'seen' by the surface of a body (such as a corpse) is called '**sky view**'. An example of sky view for a surface parallel to the ground in a wooded area is illustrated in Figure 4.4.

A limited sky view will both restrict the receipt of solar radiation and **emission** of thermal radiation during the day and the emission of thermal **earth radiation** at night (Figure 4.1). During the day the solar and sky thermal radiation not reaching the ground surface will be heating the vegetation and buildings in the view.

The sky view defines what fraction of the thermal radiation received by the surface is coming from the sky and coming from objects on the ground. The sky thermal radiation depends partly on the **effective temperature** of the air as seen from the ground. During the entire year, the western half of the US has an effective temperature of 25 °F to 43 °F cooler than the air near the surface.[17] In the eastern half of the US, the effective sky temperature is 14 to 18 °F cooler than the air temperature near the surface during the summer and 18 to 25 °F cooler during the winter.[17] Vegetation and buildings that reduce the sky view will themselves contribute thermal radiation to the ground depending in part on their surface temperature and portion of the sky view obstructed by the vegetation or building. The surface temperature of these buildings or vegetation will themselves depend on their energy balance and be influenced

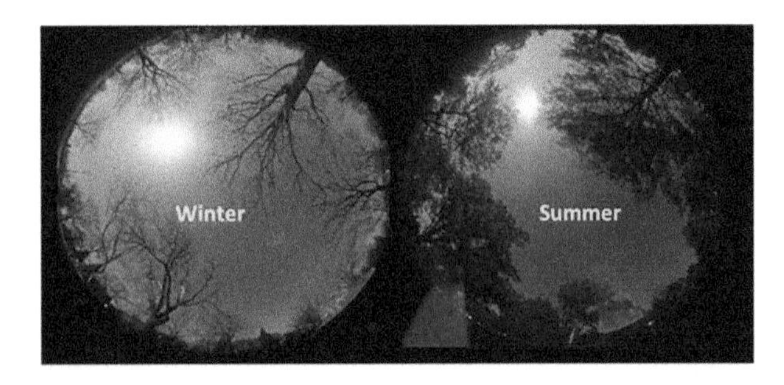

Figure 4.4 Hemispherical photos showing winter and summer sky views in an urban park. Winter sky view is 70%. Summer sky view is 47%.

Photographs by author R.H. Grant.

by their individual sky views. Since objects near the ground will have temperatures either above or near the air temperature near the ground, a greater view of the sky will tend to cool the surface during the night when there is no solar radiation- resulting in lower daily minimum air temperatures near the surface. During the winter, the canopy tends to be near the near-ground air temperature while during the summer it is elevated by the solar heating.[18] The daily mean temperature of an area with reduced sky view would be expected to be lower than that in the open air during the summer and higher than the open air in the winter.

Sky view strongly influences air temperatures in a forested area. Forested land with limited sky view has long-term average daily temperatures 1 to 5 °F cooler than nearby **open terrain**.[19–21] The average daily maximum temperatures over 30 years in nine deciduous and mixed (deciduous and conifer) forests were as much as 7 °F lower than those in open fields at similar altitudes.[21] The corresponding average daily minimum temperature from climate stations in deciduous and mixed forests was 0.7 °F above those in the open while that for the five coniferous forest locations was 3.6 °F above those in the open.[21]

The influence of sky view on urban air temperatures varies. Primarily, sky view influences daily minimum temperatures.[22] The night time temperatures in an urban 'canyon' of buildings with sky view of around 40% in Sweden was 3.6 °F warmer than in nearby open areas (sky view >96%.[23] Another study however showed very infrequent nighttime air temperature differences of 1 °F or more.[24] At some locations decreased sky view increased both night time and daytime temperatures.[25] Sky conditions affect the influence of sky view in residential areas: on clear sky days, the air temperatures were cooler than the surrounding rural areas while on cloudy days the opposite was found.[26] Clearly this highly variable influence of sky view on air temperatures in urban areas is in part due to the overall **climate**.

Sky view also influences water temperatures of ponds and streams. Rutherford and coworkers estimated that a change in relative shade of 40% to 70% would be expected to change the maximum daily water temperature 7 °F in medium streams (width 3–6 ft, depth 2–6 in, flow speed up to 0.2 mph).[27]

While forests and woodlands and urban areas have significant obstruction of the sky view at heights well above a corpse lying on the ground or in a container on the ground, grasslands have the majority of the vegetation in the bottom of the canopy.[28–32] As such, assessment of the sky view need to be made relative to the height of the corpse.

Emitted radiation

Remembering that everything radiates heat (thermal radiation), a surface still receives radiant heat even when no sunlight, either directly from the sun or that reflected by the sky, can reach a surface. The radiated heat is related to the temperature of the object and the composition of the surface. A small sky view will limit the radiant heat emitted from the ground transferred to the sky and that radiated to the vegetation or building obstructing the sky view. So, a body on the ground will exchange radiant heat with the ground, the air, and the vegetation/buildings in the local area. Human skin radiates 98% of the possible radiant heat associated with its surface temperature.[17] A nude corpse radiates 200 to 600 BTU in the first hour, and around 100 BTU per hour for the next six to 18 hours.[33]

Surface temperatures

Radiated heat depends in part on the surface temperature. Surface temperatures generally differ from that of the air over the surface since the heat capacity of dry soil (9 to 21 BTU/(ft^3 °F)[34]) is so much higher than that of the air (Table 4.1). However, the surface and air temperatures in open terrain are usually closely related so that increasing surface temperatures correspond with increasing air temperatures and decreasing surface temperatures corresponding with decreasing air temperatures. Within an oak forest, the air temperature differences between that at the surface and at 6.5 ft were not significantly different.[20] The variation in air temperatures for nine locations in an oak forest on a south-facing slope was 1.3 to 2.9 °F while that for soil surface temperatures was 2.3 to 4.5 °F.[20] The average daily surface temperature from 2950 ft to 11,000 ft **asl** across the southwestern US was 3.3 °F greater than air temperature.[35]

Exchanging heat by conduction

Conduction is the transfer of sensible heat through a solid by the collision of molecules. The ability to conduct heat into a solid is termed the '**thermal conductivity**' of the solid material (Tables 4.1 and 4.2). The conductivity describes the ability to transfer heat over an hour across a one-inch thick material of unit area given a one degree gradient in temperature (BTU in/(hr ft^2 °F). High conductivities means that heat is readily transferred through the material given a temperature gradient. Material with a low density of molecules, such as air, has a much lower conductivity than a material with a higher density of molecules such as water or stone. Because the conductivity of water is relatively high, when a material is wet or largely permeated by water, the conductivity of the material is higher than when it is dry: heat is more readily transferred through water and wet materials like wet forest litter or silt-loam soils than the corresponding dry materials.

Heat is conducted into the ground from a solid surface when there is more energy received by the surface from the sun and atmosphere than is

Table 4.1 Thermal properties of natural materials

Material	Heat capacity (BTU/(ft^3 °F)	Thermal conductivity (BTU in/(hr ft^2 °F)	Source
Dry silt-loam soil	20	0.7	36
Wet silt loam soil	52	7.3	36
Dry forest **litter**	6	0.3	37
Wet forest litter	24	1.7	37
Air (still)	0.02	0.2	34
Water (still)	62	3.9	34
Snow	3	0.6	34
Live leaves	18	2.1	38,39
Stone	34	15.1	34
Skin		2.4	40
Human body	53	3.5	3

Table 4.2 Thermal properties of building materials

Material	Heat capacity (BTU/(ft^3 °F)	Thermal conductivity (BTU in/(hr ft^2 °F)	Source
Concrete	32	10.5	34
Brick	20	5.8	34
Softwood	7	1.0	34
Hardwood	23	1.3	34
Plaster	21	2.3	34
Gypsum Board	22	1.2	34
Cellulose insulation	1	0.3	41

transferred back into the atmosphere by radiation, convection, or evaporation. Since the primary source of heat at the soil surface is usually sunlight, the periodic daily change in temperature above ground is also evident below ground. The daily range in soil temperatures (maximum to minimum) is usually less than that in the air and temporally lags the air temperature. This lag, influenced both by the heat capacity and conductivity of the soil, increases with depth in the soil.

When the ability to transfer sensible heat through a solid is low, the conductivity of the material is low (Table 4.2) and we say that the material is an insulator. Insulating materials in a house reduces heat flow both into the house and out of the house. Winter clothing is designed to be insulation. When heat flow away from an object is limited, the heat content of the object does not change rapidly and the temperature of the object stays similar over time. Insulation in buildings intends to limit heat loss through conduction through the ceiling and walls. Note that the thermal conductivity of cellulose insulation is much less than that of wood, plaster or gypsum board (Table 4.2). If two solids are in contact and at different temperatures, the proportion of heat transferred between the two solids depends on the heat capacity and conductivity. Materials with higher thermal conductivities will gain heat from the material with a lower thermal conductivity. The conductivity of walls, floors, ceilings and roofs in part define the thermal environment within buildings (Table 4.2). In conditioned buildings with Heating, **Ventilation** and Air Conditioning (HVAC), the loss of heat occurs both due to the heating of the material in the house and the materials the house is compose of, but also conduction of heat through the walls, floors, and ceilings.

The heat capacity of a corpse and water are both high. However, the conductivity of water is higher than skin (Table 4.1), a corpse will lose heat by conduction more readily if located in a water body than if located on vegetated or snow-covered land. Furthermore, the conduction heat loss to a water body is over the entire corpse surface area in contact with the water while conduction heat losses to ground is only the portion of the body that is in direct contact with the ground- typically less than 50% of the body surface area.

Exchanging heat by convection

Convection is the transfer of sensible heat through the movement of air or water. Since air and water have temperatures, they have stored heat. When the air or water moves, the heat stored in the air or water then moves with it. The moving air or water mixes with the surroundings. As air or water warms, it becomes less dense and will tend to rise regardless of whether there is a wind or current. We call this air or water '**buoyant**' and the tendency for the air or water to rise '**buoyancy**'. One can think of this rising air or water as a balloon with no walls. We consider the upward/ downward movement of air associated with the buoyancy of the air to be called '**natural convection**'. As the air or water rises it mixes with the surrounding air or water. This mixing results in a change in the surrounding air or water temperature- warming if the rising air/water is warmer than the air/water layer it is entering or cooling if the rising air/water is cooler than the air/water layer it is entering. The magnitude of natural convection depends on the temperature difference between the air and the air directly above, and is described as convection in 'still' air.

When air moves laterally, we call the transfer '**forced convection**' or '**advection**'. It is 'forced' in the sense that the heat transfer is being forced along by the wind. Winds and currents result in advection of heat as the air or water moves into air or water that is at a different temperature. Advection then is going to be closely related to the speed of the air or water. If a corpse is found in a building, the flow of air through the windows and doors, termed '**ventilation**', is an advection of heat. This ventilation, however, is a result of both temperature gradients between the inside and outside of the building but also any **atmospheric pressure** differences between air inside and outside.[42]

Convection in air

Air temperature influences the density of the air. As air warms, it expands and consequently becomes less dense. Conversely if air cools it contracts and becomes more dense. Since the pull of gravity on the air is greatest at the earth's surface, the densest air and highest atmospheric pressure is found at the surface. As we increase in altitude, the density of the air decreases and the atmospheric pressure decreases. The heating of the ground by the sun and sky is far greater than that of the air or clouds in the sky. So, as we look from ground to sky, the air near the ground is warmest and densest.

Convective heat transfer is strongly controlled by the mixing of the air from the surface to the air over the surface from horizontal wind, and upward and downward movements of the air due to changes in the air density with temperature. If the air temperature decreases rapidly with height (more than 0.55 °F per 100 ft), air pushed over a hill will want to continue to rise at the hill top (a tendency to vertically mix). If the air temperature decreases less than 0.55 °F per 100 ft, air pushed upwards by a hill will not continue to rise and mix but will descend down the back side of the hill (a tendency for vertical mixing to be suppressed). This change in temperature with altitude is called a '**lapse rate**'.

Winds result from differences in air density for a given altitude. Increased air density corresponds with increased atmospheric pressure. Air moves horizontally from high pressure to low pressure. As wind blows, it mixes the air vertically and horizontally. This mixing increases the 'conductivity' of the air over 1000 times.[34] So, heat is transferred between the surface and the air very effectively- much more effectively than it is transferred to the ground (Table 4.1).

If the wind is blowing, the air temperature from where the winds come from is also important. We describe wind directions in terms of where the wind is coming from, so typically northerly winds are colder than southerly winds in the northern hemisphere. Since winds result from differences in heating of the air, winds are calmer in the night time when the sun is not heating the ground and air.

Although the air generally is in motion, there are periods of time when the air is still. This is more common during the night time when there is no sunlight heating the ground unevenly than during the day when differences in terrain result in differences in the ground surface and air temperatures. When the ground surface is heated by solar and thermal radiation, the distribution of that heat between the air and the ground is partially dependent on the heat capacities of the air and ground and partly on the mechanisms for heat transfer (natural and forced convection for the air, conduction for the ground) because it is partially controlled by temperature gradients. For sandy soils, more heat is absorbed by the ground than goes into the air. We experience this stored heat when we walk barefoot on the beach on a sunny day- the sand is very hot and stays hot. This is because the sand has a high heat capacity and even as our foot contacts it and conducts heat from the sand, the sand still has a lot of heat. Walking on dead grass however does not feel as hot because there is very little heat in the dead grass at the same temperature and our foot rapidly takes the little heat in the grass without significantly changing our foot temperature.

The more heat stored in the soil, the longer the temperature gradient between the soil and the air is maintained and heat is exchanged depending on the magnitudes of the conductivity and the **convective exchange coefficient**. The more organic matter is in the soil, the less heat is stored in the soil, the lower the soil conductivity, and more of the heat at the ground surface is convected into the air than conducted into the ground. So high organic matter soils will increase and decrease in temperature more quickly than low organic soils because they have lower heat capacities (Table 4.1).

Winds decrease in speed as the ground is approached, so wind speeds reported by climates stations measured at 10 ft **agl** (above ground level) greatly overestimate winds at corpse height in grass, under shrubs or under trees. Wind speeds at corpse height in the open will often be less than 2 mph.[9, 43, 44] Wind speeds in forest gaps ranging from 430 ft^2 to 3078 ft^2 are less than 1 mph higher than found in a nearby forest.[45] Wind speeds at 3 ft agl in various low density 65 to 115 ft canopy forests were about 10% of that measured at 33 ft agl in the open while winds3 ft agl in two low density scrublands 7 ft to 13 ft high were about 25% of that in the open.[46] Moon and coworkers found very light winds (less than 1 mph) in many forests even when winds in the open were much higher.[46] The convective exchange coefficient of heat at these low wind speeds would be about 1.4 BTU hr^{-1} ft^{-2} $°F^{-1}$.[47–49]

A given temperature gradient will exchange more heat by conduction than convection. The heat is exchanged across a surface (hence the units of ft^2 in the convective exchange coefficient above and conductivities in Tables 4.1 and 4.2) Similarly, if the heat exchanged with the air is greater than that exchanged with the ground where a corpse is discovered, the corpse approaches the averaged air temperature rather than the ground temperature.

If a corpse surface is warmer than the air above it, heat will be lost from the corpse as it convects from the corpse. If there is little wind the heat transferred to/from the air from/to a prone corpse surface is by natural convection. A representative convective heat exchange coefficient for a body in still air is approximately 1 BTU hr^{-1} ft^{-2} $°F^{-1}$.[3] As a result, even when there is no wind, the exchange of heat from a corpse to cooler air is greater than that exchanged with the ground under the body since that ground is shaded (not receiving solar radiation heating). Since snow is a good insulator (Table 4.1), a corpse lying on snow during the winter will exchange heat to the air more than with the snow-covered ground. Increased wind speeds amplify the transfer of heat given the same difference in temperatures. In enclosed spaces such as buildings when the air and walls of similar temperature, the natural convection of sensible heat will be greater than the radiant heat exchanged with the environment.[50]

Convection in water

The temperature of a stream or river water at any specific spot is primarily influenced by the energy budget of the upstream conditions in the watershed. The stream or river temperatures are generally close to that of the groundwater at the head of the river or stream.[51] This is because the heat capacity of the water is high, so changes in water temperature are slow. If a stream is flowing at five mph, the water at some instant in time where a body is found has traversed 5 miles in the prior hour. Rutherford and coworkers found that it required four hours of flow (traveling 3/4ths of a mile) for a stream temperature to adjust to a change in the energy budget for second order streams (streams in width 3–6 ft, depth 2–6 in, flow speed up to 0.2 mph).[27] Daily maximum temperatures in streams changed by 7 °F within 0.4 to 0.9 miles of changes in vegetation cover of the same second-order stream.[27]

The next most important factor in the stream or river temperature is the **net radiation.**[52] This results in distinct variations in temperature over the course of a day with greater variation in small streams than large rivers. Sky view over the stream or river is therefore important in estimating the water temperature, as it is in estimating the air temperature.

Given near- constant sky views, the changes in the net radiation over days due to cloud cover, **precipitation** and winds also result in changes in water temperature. Ham and coworkers found water temperature to vary by an average of 7.6 °F in streams and small ponds (1–3 acre) of a 57-acre suburban watershed during the summer.[53] The daily minimum and maximum stream water temperatures lag (or happens after) the air temperatures. The average time lag between the daily air temperature cycle and the daily water temperature cycle for thirteen rivers in Minnesota, Louisiana, Arkansas and Texas was four hours for clear sky days.[7]

Stream temperatures vary less when the water flow is greater. The greatest flow rate in a river is often on the surface in the center of the river (Figure 4.5). If the corpse is trapped by debris in the river, the flow rate is probably somewhat less than the maximum. If the corpse is found while still moving down the river, the difference between how fast the corpse is moving and the water around the corpse is moving defines the 'speed of the flow' relative to the corpse, and not the speed of the water itself.

Heat exchange between a body in water is more than 15 times that for a body in still air.[3, 54] The addition of water flowing across a body increases the heat exchange because the water that was heated by the body has moved

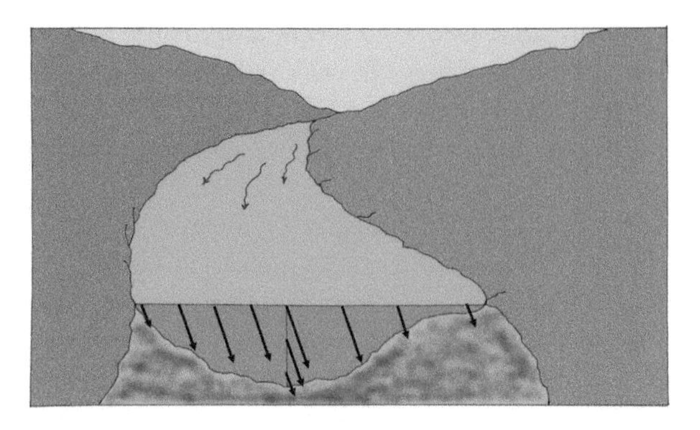

Figure 4.5 Variation in water flow speed with depth and span across a river. Flow rate indicated by length of arrows.

Illustration by E.F. Grant.

away from the body. A representative convective heat exchange coefficient for still water is approximately 18 BTU hr^{-1} ft^{-2} $°F^{-1}$ while that for water moving at 0.6 mph is 176 BTU hr^{-1} ft^{-2} $°F^{-1}$.[3]

Local temperature climates

There are usually hour-to-hour changes in air temperature at the location of a corpse as a result of changes in the energy balance. We have already looked at the causes for changes in air temperature over flat land over a day (Figure 4.2). But much of the land surface is not flat and much of it has vegetation of some sort or has buildings on it. These changes in the terrain influence the air temperature over a corpse lying in the terrain. Changes in terrain elevation, roughness and/or material often influence the surface and air temperatures and create winds. Of these changes, there are a number of winds that influence small 'local' areas over time periods of hours to days including: mountain-valley winds, lake-land winds, urban-rural winds and downslope winds. The overall impact of the various local winds on the local climate are most evident when the winds associated with the overall **weather** patterns are relatively steady and light, as under high pressure systems. Over longer monthly or annual periods, the frequency of weather systems with associated cloudiness, winds, and precipitation influences the local climate.

Due to elevation

Air changes temperature when it rises or descends so air in the mountains will usually be colder than on the lower altitude plains. When air descends, it compresses and warms. This effect of compression is evident in the temperature of pumps and compressors that put air into your car or bicycle tire. When air rises at expands and cools. The cooling on expansion is evident in the change in temperature of a can of carbonated beverage when you open the can. The tendency for the air to mix is related to the rate of decrease in air temperature with altitude. A rate of decrease in air temperature on rising of −0.55 °F per 100 ft is called the '**adiabatic** lapse rate' (the cooling that air will experience if it is lifted 100 ft due just to the expansion of the air volume due to reduction in atmospheric pressure above that height). It will also warm at that rate if the air descends and compresses. This describes the change in temperature with altitude when the air is well mixed.

The rising and sinking air is however influenced by winds. Changes in the temperature and moisture of air advected to an area however strongly influences the observed lapse rates.[55] Lapse rates on slopes and in valleys influenced by local winds such as the mountain-valley and Foehn and Chinook winds also vary widely.[56] Clearly the daily and seasonal variation in observed lapse rates with local conditions shows that the 'adiabatic lapse rate', while theoretically sound, does not account for influences of the earth's surface on the air above the ground. A 'standard' atmosphere was developed to represent the average rate of change of temperature with height. This standard atmosphere lapse rate is −0.35 °F per 100 ft up to an altitude of 36,000 ft.[57]

Lapse rates varied with season and for maximum and minimum daily temperatures in the Cascade mountains- smallest in late summer minimum temperatures and highest in the maximum temperatures.[58] Lapse rates of the average daily temperature for 30 stations over 30 years across the Waterton-Glacier National Park averaged −0.37 °F per 100 ft of altitude.[19] Precipitation and the presence of snow on the ground also influences the lapse rate. The lapse rate over the period of year on the eastern side of the Rocky Mountain Range assessed for Case 8 in the Appendices averaged −0.51 °F per 100 ft of altitude.

Due to advection in hilly or mountainous terrain

Hilly or mountainous terrain create a range of local climates depending on the steepness and orientation of the slopes. Since the angle between the sun beam and the surface strongly influences the received solar radiation, the varying steepness of the slope and orientation of the slope result in varying solar radiation received at any given hour. On sloped terrain, the changing position in the sky of the sun during the day changes the orientation of the sun to the ground and how much solar radiation is received by the ground. Considering now the thermal radiation from the ground and sky: the smaller sky view in a valley than mountain (hill) top results in differences in incoming solar radiation and outgoing thermal radiation over the course of the day.

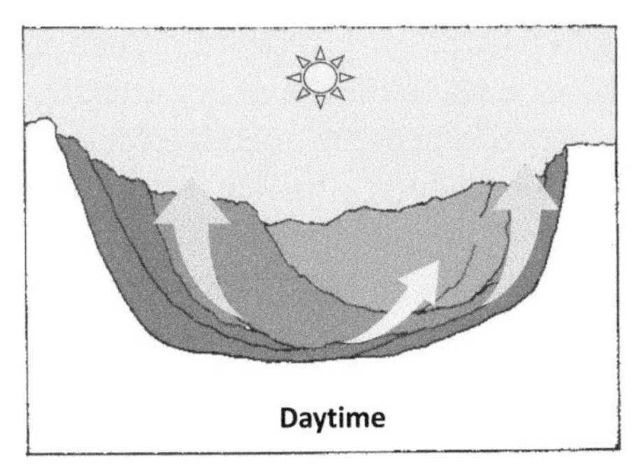

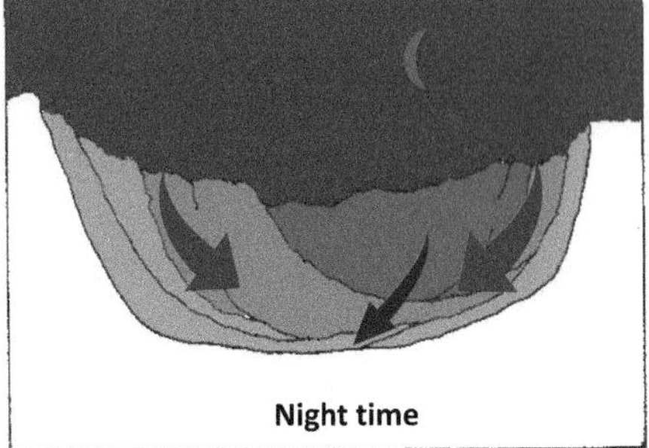

Figure 4.6 Air flow during day and night in a valley/slope terrain.

Illustration by E.F. Grant.

Since solar and sky radiation are the dominant heat sources for the earth's surface, the heating of the surface and the air directly above the surface varies widely. As the day progresses the heating of the peak and slopes increases more quickly than the valley and the warmed air is **buoyant** and rises up the slope removing air by advection in the valley (Figure 4.6).

The strongest winds upslope in the northern hemisphere occurs on slopes facing south,[59] corresponding to the slope with the greatest solar heating. During the night, the air along the peaks cools fastest due to the greater sky view than the valley. As the air cools it becomes more dense and descends the slope and pools in the valley (Figure 4.6). If the valley has an outlet onto a plain, the cold air continues to flow out onto the plain. The downslope wind speeds do not vary with slope orientation.[59]

The long-term average daily temperatures in hilly terrain in Ohio showed temperature on a ridge were 10 °F to 12 °F warmer than those near the top of a valley throughout the year.[60]

Due to proximity to water bodies

Water bodies change local climates. Since water has a higher heat capacity than land (Table 4.1) the daily heating and cooling of the land is greater than water bodies. This creates higher air temperatures over the land than over the water body during the day and higher air temperatures over the water body than the land during the night. During the day, the high land temperatures cause the air to become buoyant and rise. This air is replaced at the surface by the cooler more-dense air over the water resulting in a lake/sea breeze.

The influence of a water body on the lake air temperature of the air blowing from the water to the land depends partly on the breadth of the water body. A distinct lake breeze is not likely to be evident for lakes less than one mile across.[61] The distance of influence of the water body across the land is also limited. The advection of a cool breeze extended 12 mi to 30 mi inland of the ocean in the mid-**latitudes**.[59] Lakes breezes extending 8 to 31 miles inland have been reported for lakes between two and 20 miles across.[61] Segal and coworkers found that a lake breeze will penetrate inland less than one-half the width of the lake.[61]

The long-term average daily temperature influence of lakes on surround landscape extends approximately 50 mi from the Great Lakes (53 mi to 183 mi across) in the US.[62] Daily minimum temperatures are increased during the entire year by proximity to the Great Lakes. The increase in winter minimum temperatures within the 50 mi band is about 2 °F upwind of the Lakes and 2° to 4 °F downwind.[62] During the summer, the average daily minimum temperatures with the influence zone is between an increase of 2 °F upwind to a decrease of 2 °F downwind.[62] Daily maximum temperatures are reduced by the Lakes during Spring and Summer and slightly increased during Fall and Winter.[62] In general, the proximity to the lakes decreases the daily maximum temperatures up to 4 °F upwind and downwind near the shore. Because the variations are strongly influenced by the solar heating, the average daily temperature can still be well-estimated by a daily maximum and minimum temperature **record**.

Due to advection near urban areas

The varying density of population across the country influences the variability in terrain surface temperature and local climates. Since man-made materials typically have higher heat capacities than vegetated land and urban landscapes vary in height more than most vegetated areas, more heat is stored within urban areas than vegetated rural areas. During the sunny daytime the air in urban areas heats up more quickly than the rural areas but the maximum daily air temperatures around 4 pm is similar (Figure 4.7). During the night the urban area cools more slowly than the rural area and the daily minimum temperature is greater in the urban area than the surrounding rural landscape (Figure 4.7). Air temperatures in the afternoon are higher in the city than in the rural surroundings, for example it is 2 °F to 4 °F warmer for a city in France.[63]

A rural-urban breeze is driven by this temperature gradient. The breeze reduces the effects of sky view on night time cooling in the urban area.[23] Since rural areas typically ring urban areas, the rural breeze converges on the urban area and enhances the lifting of the air over the urban center. The breeze is generally very light[64] and may be detectable only when the entire area is under relatively calm wind conditions such as under high pressure, however the breeze may reduce or increase speed within the urban area depending on many factors.[34]

Overall, urbanization appears to influence the minimum daily air temperature much more than the maximum temperature.[65] Enhanced air temperatures in urban areas appear to be greatest in fall evenings.[66] The long-term

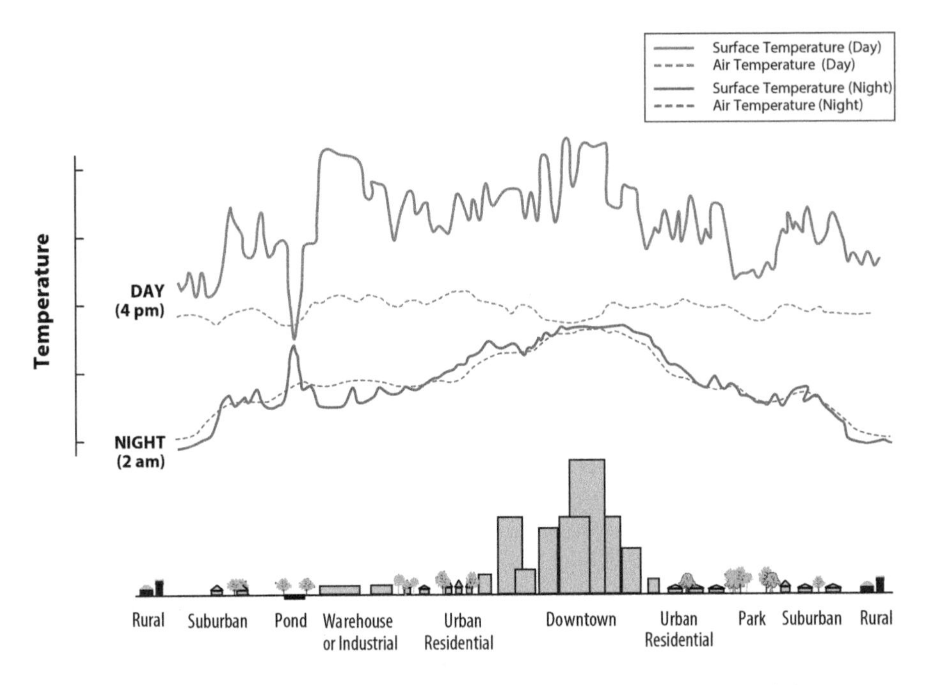

Figure 4.7 Typical variation in maximum (day) and minimum (night) temperatures across an urban transect.

Source: USEPA (https://www.epa.gov/heatislands/learn-about-heat-islands)

average minimum daily temperatures within 21 cities over three years in the USA was 0.7 °F higher than the surrounding rural areas while the urbanization effect for the daily mean temperature was 0.5 °F.[66] The urban 'heat island' results in an average 4 °F higher minimum night time temperature for climates with average annual temperatures of about 72 °F.[67] There are some cities where the urban area is cooler than the surrounding rural area- cities in dry climates without surrounding rural vegetation such as Phoenix Arizona. Manoli and coworkers found that this occurs for climates with average temperatures greater than about 77 °F and less than about 31 inches of rain per year.[67]

Since cities are often located at coastlines or in valleys, the interactions of the night time rural breeze driven by the urban heat storage from the daytime under near-calm conditions may be influenced by downslope and down valley winds.

Due to advection of downslope winds

Downslope winds in mountainous terrain affect the local climates at the base of the mountains. Since winds generally move from west to east in the US due to global circulation patterns, the downslope winds are usually found to the

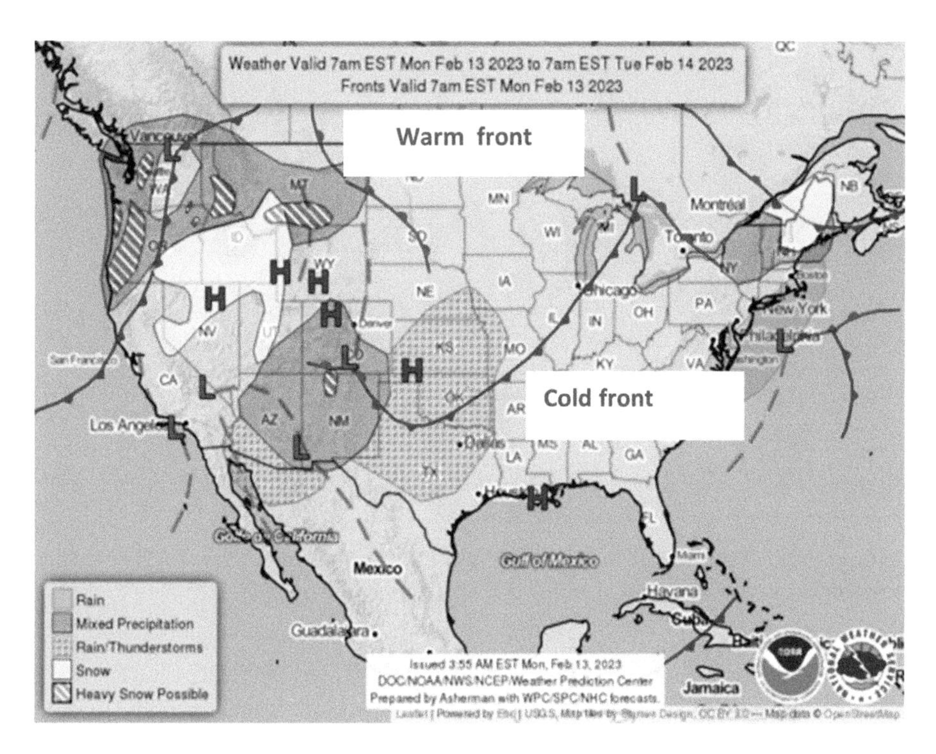

Figure 4.8 National Weather Service **Synoptic weather** forecast map showing Fronts and high (H) and low (L) pressure centers.

Source: USA Weather Prediction Center Archives, National Center for Environmental Prediction, NOAA; www.wpc.ncep.noaa.gov/noaa/noaa_archive

east of north-south oriented mountain ranges typically found in the western US and to the south of east-west oriented mountain ranges in Europe. If the area is under high pressure (Wyoming in Figure 4.8), the wind blowing over a mountain range creates waves downwind (often evident in parallel cloud patterns) and moves rapidly down the lee side of the mountain.[59, 68]

Two major types of downslope winds occur: hot and cold downslope winds. If the air upwind of the mountain is relatively warm, the downslope winds on the downwind side are hot, dry and fast. This is called a 'Chinook' or 'Foehn' wind in the US but has many different names across the world. These winds are most common during the night and during the winter: during the winter 40% of all westerly winds are Chinooks with 60% of those increase the air temperature on the downwind side of the mountain more than 10 °F within a few hours and last for more than a day.[69] Since these winds last for more than a day and are strongly influencing the air temperature, the variation in air temperature over the day is not cyclical (Figure 4.9). As a result, the average daily temperature may not be well-estimated by a daily maximum and minimum air temperature climate record.

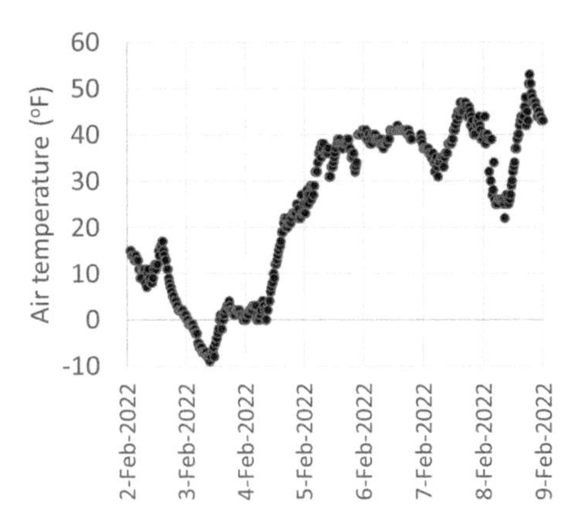

Figure 4.9 Air temperature variation during a chinook wind, Helena, MT, USA. Chinook wind occurred on 6 Feb 2023.

Data source: NOAA National Center for Environmental Information, https://www.ncei.noaa.gov/access

If the air upwind of the mountain is relatively cold and the mountains are not very high, the descending air tends to be cold, dense, and dry on the downwind side. This is called a 'Bora' or 'Sirocco' wind, occurring most frequently along the coast of Europe during night time in the winter. These winds can be over 30 mph[70] and will greatly influence near-surface air temperatures through enhanced convection of heat from the surface and advection of cold air. These winds also can last multiple days, resulting in the average daily temperature to be poorly-estimated by a daily maximum and minimum air temperature record.

Due to advection by gravity-driven winds

The local climates around hilly terrain are also influenced by winds driven by the density of the air. During calm clear nights the temperature near the surface usually increases with height. In these light wind conditions, the local terrain can strongly influence the movement of the air. As air cools during the night, a slight slope in the ground (≥ ½ degree slope) will cause air to flow downward by the pull of gravity and accumulate in depressions or 'hollows'. Density-driven winds are typically light.[71] As with water flowing down a slope, the intensity of this flow depends on the overall area that has air converging to a depression like a river basin does to a river water flow.[71] This resulted in depressions and stream channels in a rolling terrain in Australia to be up to 4 °F cooler than expected.[71] Daily maximum temperatures in the summer in a frost pocket on a northern Allegheny Mountain plateau were similar to those found in the open level ground but the daily minimum temperatures were about 5 °F cooler.[72]

The very slow movement of this air greatly limits the convection of heat from the soil to the air so the soil temperatures in the cold pocket in the northern Allegheny Mountain plateau did not greatly differ from those elsewhere in the open.[70]

Impacts of mid-latitude large-scale weather changes on air temperatures

The interval between last sighting of the person and discovery of the corpse defines the needed time **resolution** needed to estimate the amount of heat available for insect development and decomposition. It is important to assess the heat available for decomposition and insect maggot development from hourly temperatures when the time between last sighting and discovery is in the order of a week or less.

Weather systems across the middle latitudes of the earth (between about 30° latitude and 55° latitude) result in transitions of air temperatures on three-day intervals in the winter and six-day intervals in the summer in the northern parts of the US and Europe[73] to only every 19 days or more in the southern US.[74] The frequency of cloud cover also often varies with weather systems as the temperature of the air, the tendency for air to mix.

The local climates influenced by the local winds are most evident when the winds are not being influenced by large-scale weather systems. This occurs most often when the area of interest is under relatively high atmospheric pressure (see Nebraska in Figure 4.8) for multiple days. When we have relatively low barometric air pressure (see above Michigan Figure 4.8), cloud cover increases, winds are moderate to high and rain or snow is possible. Winds and rainfall can combine to influence the air temperature as cold **front**s, warm fronts, and low pressure centers pass over a location. When the winds increase, the advection of heat can change the air temperature up to around 5 °F per hour. Rainfall or snowfall change the radiative and conductive properties of the ground and hence also change the energy balance. As a result, the average daily temperature may be overestimated or underestimated from measured minimum and maximum daily temperatures depending of the timing of a frontal passage during the day.

Water temperatures are far less influenced by the passage of large-scale weather systems due to their greater heat capacity (Table 4.1). The lag in water temperatures behind the air temperatures was estimated at one to three days for 596 stream gaging stations across the US.[75] The water temperatures of eleven rivers with total flows ranging from 14 to 73070 cubic feet per year lagged the daily air temperatures by 0 to 8 days with the longest lag associated with a large river during a low volume year.[7]

References

1. Chumlea, W.C., Guo, S.S., Zeller, C.M., Reo, N.V. and Siervogel, R.M., 1999. *Kidney International*, 56(1), pp.244–252.
2. Xiang, S.H. and Liu, J., 2008. *International Journal of Heat and Mass Transfer*, *51*(23–24), pp.5486–5496.
3. Abraham, J., Cheng, L., Vallez, L. and Wei, T., 2022. *Journal of Forensic Sciences*, *67*(3), pp.1049–1059.
4. Molnar, G.W., Hurley, H.J. and Ford, R. 1969. *Pflüger's Archiv für die gesamte Physiologie des Menschen und der Tiere*, *311*, 16–24.
5. Henssage, C. 2023. pp. 77–186 In: Madea, B. (Ed) *Estimation of time since death*. 4th Ed. CRC Press, Oxford, England. 383p.
6. Woolway, R.I., Jones, I.D., Maberly, S.C., French, J.R., Livingstone, D.M., Monteith, D.T., Simpson, G.L., Thackeray, S.J., Andersen, M.R., Battarbee, R.W. and DeGasperi, C.L., 2016. *PLOS one*, *11*(3), p.e0152466.
7. Stefan, H.G. and Preud'homme, E.B., 1993. *JAWRA Journal of the American Water Resources Association*, *29*(1), pp.27–45.
8. Forster, P., T. Storelvmo, K. Armour, W. Collins, J.-L. Dufresne, D. Frame, D.J. Lunt, T. Mauritsen, M.D. Palmer, M. Watanabe, M. Wild, and H. Zhang, 2021. pp.923–1054 In Masson-Delmotte, V., P. Zhai, A. Pirani, S.L. Connors, C. Péan, S. Berger, N. Caud, Y. Chen, L. Goldfarb, M.I. Gomis, M. Huang, K. Leitzell, E. Lonnoy, J.B.R. Matthews, T.K. Maycock, T. Waterfield, O. Yelekçi, R. Yu, and B. Zhou (Eds.) Climate Change 2021: The Physical Science Basis. Contribution of Working Group I to the Sixth Assessment Report of the Intergovernmental Panel on Climate Change Cambridge University Press, Cambridge, United Kingdom and New York, NY, USA.
9. Geiger, R. 1965. *Climate near the ground*. Harvard University Press, Cambridge, MA, 611p.
10. Buffo, J., Fritschen, L.J. and Murphy, J.L. 1972. *Research Paper* PNW-RP-142. Portland, OR: U.S. Department of Agriculture, Forest Service, Pacific Northwest Research Station. 75 p.
11. Underwood, C.R. and Ward, E.J., 1966. *Ergonomics*, *9*(2), pp.155–168.
12. Goward, S.N. 1981. *Physical Geography*, *2*, 19–33.
13. Campbell, G.S. and Norman, J.M. 1988. *An introduction to environmental biophysics*. 2nd Ed., Springer-Verlag, NY. 286p.
14. Stanhill, G., Hofstede, G.J. and Kalma, J.D., 1966. *Quarterly Journal of the Royal Meteorological Society*, *92*(391), pp.128–140.
15. Anderson, R.R. and Parrish, J.A., 1981. *Journal of Investigative Dermatology*, *77*(1), pp.13–19.
16. Watanabe, S., Horikoshi, T. and Tomita, A. 2010. *Japanese Journal of Biometeorology*, *47*, 165–173.
17. Martin, M. and Berdahl, P., 1984. *Solar energy*, *33*(3–4), pp.321–336.
18. Pomeroy, J.W., Marks, D., Link, T., Ellis, C., Hardy, J., Rowlands, A. and Granger, R., 2009. *Hydrological Processes: An International Journal*, *23*(17), pp.2513–2525.
19. Finklin, A.I. 1986. *Gen. Tech. Rpt.* INT-204. Forest Service, United States Department of Agriculture, Intermountain Experiment Station, 124p.
20. Xu, M., Chen, J. and Brookshire, B.L., 1997. *Climate Research*, *8*(3), pp.209–223.

21. Renaud, V., Innes, J.L., Dobbertin, M. and Rebetez, M., 2011. *Theoretical and Applied Climatology*, *105*, pp.119–127.
22. Hu, L., Monaghan, A., Voogt, J.A. and Barlage, M., 2016. *Remote Sensing of Environment*, 181, pp.111–121.
23. Holmer, B., Thorsson, S. and Eliasson, I., 2007. *Geografiska Annaler: Series A, Physical Geography*, *89*(4), pp.237–248.
24. Eliasson, I., 1996. *Atmospheric Environment*, *30*(3), pp.379–392.
25. Ha, J., Lee, S. and Park, C., 2016. *Sustainability*, *8*(9), p.895.
26. Svensson, M.K. and Eliasson, I., 2002. Diurnal air temperatures in built-up areas in relation to urban planning. *Landscape and Urban Planning*, *61*(1), pp.37–54.
27. Rutherford, J.C., Marsh, N.A., Davies, P.M. and Bunn, S.E., 2004. *Marine and Freshwater Research*, *55*(8), pp.737–748.
28. Moss, D.N. 1965. *Agricultural Meteorology, Meteorological Monographs*, *6*(28), 90–108.
29. Fliervoet, L.M., 1987. *Vegetatio*, *70*, pp.105–117.
30. Nassiri, M., Elgersma, A. and Lantinga, E.A., 1996. pp. 269–274 In *Proc. 16th General Meeting European Grassland Federation: Grassland and land use systems, Vol. 1*, Grassland science in Europe, G. Parente et al. (eds.). Grado (Gorizia), Italy.
31. Jurik, T.W. and Kliebenstein, H., 2000. *The American Midland Naturalist*, *144*(1), pp.51–65.
32. Galzerano, L., Malheiros, E.B., Morgado, E.D.S. and Ruggieri, A.C., 2012. *Nucleus Animalium*, *4*(2), pp.11–18.
33. Oke, T.R. 1978. *Boundary layer climates*. Methuen Press, London. 372p.
34. Mall, G., Hubig, M., Beier, G. and Eisenmenger, W., 1998. *International Journal of Legal Medicine*, *111*, pp.299–304.
35. Mutiibwa, D., Strachan, S. and Albright, T., 2015. *IEEE Journal of Selected Topics in Applied Earth Observations and Remote Sensing*, *8*(10), pp.4762–4774.
36. Ochsner, T.E., Horton, R. and Ren, T., 2001. *Soil science society of America Journal*, *65*(6), pp.1641–1647.
37. Riha, S.J., McInnes, K.J., Childs, S.W. and Campbell, G.S., 1980. *Soil Science Society of America Journal*, *44*(6), pp.1323–1325.
38. Hays, R.L., 1975. *Planta*, *125*, pp.281–287.
39. Jayalakshmy, M.S. and Philip, J., 2010. *International Journal of Thermophysics*, *31*, pp.2295–2304.
40. Kuroda, F., Hiraiwa, K., Oshida, S. and Akaishi, S., 1982. *Medicine, Science and the Law*, *22*(4), pp.285–289.
41. Kumar, D., Alam, M., Zou, P.X., Sanjayan, J.G. and Memon, R.A., 2020. *Renewable and Sustainable Energy Reviews*, *131*, p.110038.
42. ASHRAE. 2001. *2001 Fundamentals, SI edition*. Amer. Soc. Heating, Refrig., and Air Condit. Engineers, Inc, Atlanta GA.
43. Waterhouse, F.L., 1955. *Quarterly Journal of the Royal Meteorological Society*, *81*(347), pp.63–71.
44. Launiainen, S., Vesala, T., Mölder, M., Mammarella, I., Smolander, S., Rannik, Ü., Kolari, P., Hari, P., Lindroth, A. and Katul, G., 2007. *Tellus B: Chemical and Physical Meteorology*, *59*(5), pp.919–936.
45. Abd Latif, Z. and Blackburn, G.A., 2010. *International Journal of Biometeorology*, *54*, pp.119–129.

46. Moon, K., Duff, T.J. and Tolhurst, K.G., 2019. *Fire Safety Journal, 105*, pp.320–329.
47. Ishigaki, H., Horikoshi, T., Uematsu, T., Sahashi, M., Tsuchikawa, T., Mochida, T., Hieda, T., Isoda, N. and Kubo, H., 1993. *Journal of Thermal Biology, 18*(5–6), pp.455–458.
48. Quintela, D., Gaspar, A. and Borges, C., 2004. *European Journal of Applied Physiology, 92*, pp.663–668.
49. Mao, N., Song, M., Pan, D. and Deng, S., 2017. *Applied Thermal Engineering, 117*, pp.385–396.
50. Kurazumi, Y., Tsuchikawa, T., Ishii, J., Fukagawa, K., Yamato, Y. and Matsubara, N., 2008. *Building and Environment, 43*(12), pp.2142–2153.
51. Caissie, D., 2006. *Freshwater Biology, 51*(8), pp.1389–1406.
52. Webb, B.W. and Zhang, Y., 1997. *Hydrological Processes, 11*(1), pp.79–101.
53. Ham, J., Toran, L. and Cruz, J., 2006. *Environmental Geology, 50*, pp.55–61.
54. Boutelier, C., Bougues, L. and Timbal, J., 1977. *Journal of Applied Physiology, 42*(1), pp.93–100.
55. Blandford, T., K. Humes, B. Harshburger, B. Moore, V. Walden, and H. Ye. 2008. *Journal Applied Meteorology and Climatology, 47*(1), pp.249–261.
56. Rolland, C. 2003. *Journal of Climate, 16*(7), pp.1032–1046.
57. NASA. 1976. *U.S. Standard atmosphere*, 1976. National Oceanic and Atmospheric Administration, National Aeronautics and Space Administration and United States Air Force. NASA-TM-X-74335. 241p.
58. Minder, J.R., Mote, P.W. and Lundquist, J.D., 2010. *Journal of Geophysical Research: Atmospheres, 115*(D14122).
59. Atkinson, B.W. 1981. *Meso-scale Atmospheric circulations*. Academic Press, London. 495p.
60. Wolf, J.N., Wareham, R.T., and Scofield, H.T., 1949. *Bulletin, Ohio Biological Survey*, Bulletin 41, 267p.
61. Segal, M., Leuthold, M., Arritt, R.W., Anderson, C., Shen, J. 1997. *Bulletin, American Meteorological Society, 78*, pp.1135–1147.
62. Scott, R.W. and Huff, F.A., 1996. *Journal of Great Lakes Research, 22*(4), pp.845–863.
63. Hidalgo, J., Pigeon, G. and Masson, V., 2008. *Meteorology and Atmospheric Physics, 102*(3–4), pp.223–241.
64. Wong, K.K. and Dirks, R.A., 1978. *Journal of Applied Meteorology and Climatology, 17*(5), pp.677–688.
65. Heisler, G.M. and Brazel, A.J., 2010. *Urban Ecosystem Ecology, 55*, pp.29–56.
66. Epperson, D.L., Davis, J.M., Bloomfield, P., Karl, T.R., McNab, A.L. and Gallo, K.P., 1995. *Journal of Applied Meteorology and Climatology, 34*(2), pp.358–370.
67. Manoli, G., Fatichi, S., Schläpfer, M., Yu, K., Crowther, T.W., Meili, N., Burlando, P., Katul, G.G. and Bou-Zeid, E., 2019. *Nature, 573*(7772), pp.55–60.
68. Grisogono, B. and BelusŠIĆ, D., 2009. *Tellus A: Dynamic Meteorology and Oceanography, 61*(1), pp.1–16.
69. Ives, R. L. 1950. *Annals of the Association of American Geographers, 40*(4), 293–327.
70. Belušić Vozila, A., Belušić, D., Telišman Prtenjak, M. *et al.* 2023. *Climate Dynamics* https://doi.org/10.1007/s00382-023-06703-z

71. Laughlin, G.P. and Kalma, J.D., 1987. *Agricultural and Forest Meteorology*, *40*(1), pp.1–16.
72. Hough, A.F., 1945. *Ecology*, *26*(3), pp.235–250.
73. Wang, X.L., Feng, Y., Compo, G.P., Swail, V.R., Zwiers, F.W., Allan, R.J. and Sardeshmukh, P.D., 2013. *Climate Dynamics*, *40*, pp.2775–2800.
74. Mitchell, T.J., Knapp, P.A. and Ortegren, J.T., 2022. *Meteorology*, *1*(2), pp.211–219.
75. Bogan, T., Mohseni, O. and Stefan, H.G., 2003. *Water Resources Research*, *39*(9), 1245.

Controls on local climate humidity

5

The effect the local humidity climate on a body

The **humidity** of our local climate influences the moisture content of a body in the environment. The water content of the male and female human body is about 51% and 44%.[1] We feel hotter in air with high humidity than low humidity because our body removes core heat produced by metabolism through liquid water **evaporation** and **sensible heat convection** in your breath and from skin surface. Since corpse decomposition depends in part on the availability of water for microbial activity, the loss of water from a corpse requires a knowledge of the humidity of the environment where the corpse is found.[2] If the corpse is above ground, the humidity of the air is used to assess potential for microbial activity. If the corpse is buried, **soil pore** humidity as well as soil **temperature** is used to assess potential for microbial activity.

Water found either in environment is in vapor, liquid, or solid form. The amount of vapor, liquid, and solid water at a location depends on a balance of sources and losses just as the temperature depends on the balance of thermal energy sources and losses discussed in Chapter 4.

Measures of humidity

Air typically contains between 1% and 3% water as a gas (**water vapor**). There are several ways to describe how much water is in the air. When we think about humidity, we almost always think **relative humidity**, or **RH**. This is a measure that goes from zero to 100%. We know that when it is dry out the RH is low and when it is raining or **fog**gy out the RH is about 100%. But what is 'relative' about RH? Based on the scale of zero to 100%, we can see that zero means there is no RH. This means there is no moisture in the air- an unlikely situation in the natural or man-made environment. But what does 100% RH mean? At 100% RH the air cannot hold any more water- it is saturated with water vapor. When it saturates, it usually condenses and appears as small water droplets in the air- fog or clouds.

DOI: 10.4324/9781003486633-5

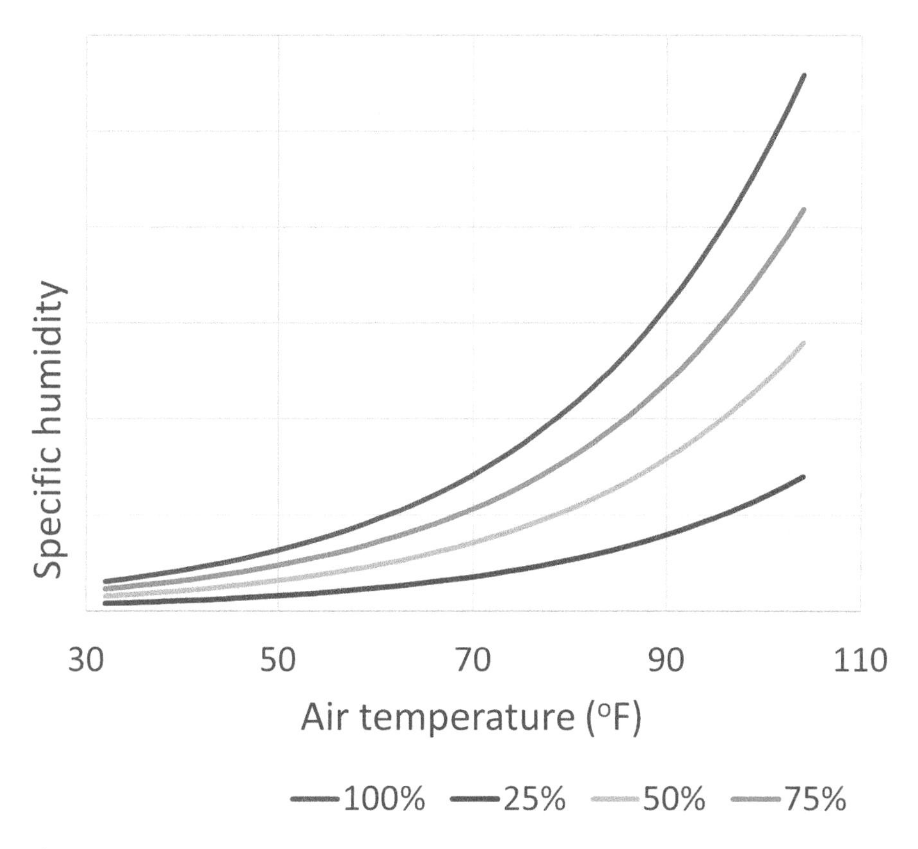

Figure 5.1 Variation in relative humidity, specific humidity with air temperature.

In our atmosphere, the amount of water vapor that the air can hold varies with air temperature in a regular way (Figure 5.1). If we describe how much water the air can hold at a given temperature in terms of the amount of water vapor in an amount of air (such as pounds of water per pound of air), we can illustrate this relationship as in Figure 5.1. This second measure is called **specific humidity**. The specific humidity at saturation for a given temperature is indicated by the blue line in Figure 5.1. For the same air temperature, a 50% RH means that the air has 50% of the specific humidity at saturation (yellow line in Figure 5.1). The set of colored RH curves in Figure 5.1 show the different amount of specific humidity (vertical axis) of the air given an air temperature (horizontal axis). RH is commonly measured by automated **weather** stations used by private companies, state networks and individuals. It is also commonly measured by handheld instruments.

The humidity of the air can also be described by the air temperature and the **dew point temperature**. The dew point temperature is the temperature of air when **condensation** occurs as you cool the air without changing the specific humidity of the air (moving left to right in Figure 5.1). Such condensation happens when you turn on the air conditioning in your car on a humid summer day-- the cold air cools the windshield to the dew point temperature of the air outside the car. The dew point temperature is measured at airports using the **Automated Surface Observation System (ASOS)**.

The humidity of the air can also be described by the air temperature and the **wet bulb temperature**. The wet bulb temperature is a temperature representing the sensible heat equilibrium between the cooling of a wet surface that is evaporating liquid water into water vapor in the air and the heating of that surface as water vapor condenses onto the wet surface. To get this equilibrium, the air needs to be rapidly moving over the wet surface to maximize the convection of sensible heat and moisture. This is most commonly measured by using hand-held psychrometers to evaluate indoor humidity levels.

Daily variation of humidity in the air

Since RH is a relative measure and air at different temperatures can hold different amounts of water, the RH over the course of a typical day will vary (Figure 5.1). In general, the RH varies inversely to the air temperature on clear days, with maximum RH and minimum air temperatures during the night and minimum RH and maximum air temperatures during the day (Figure 5.2). If clouds form during the day, the reduced **solar radiation** (Chapter 4) causes

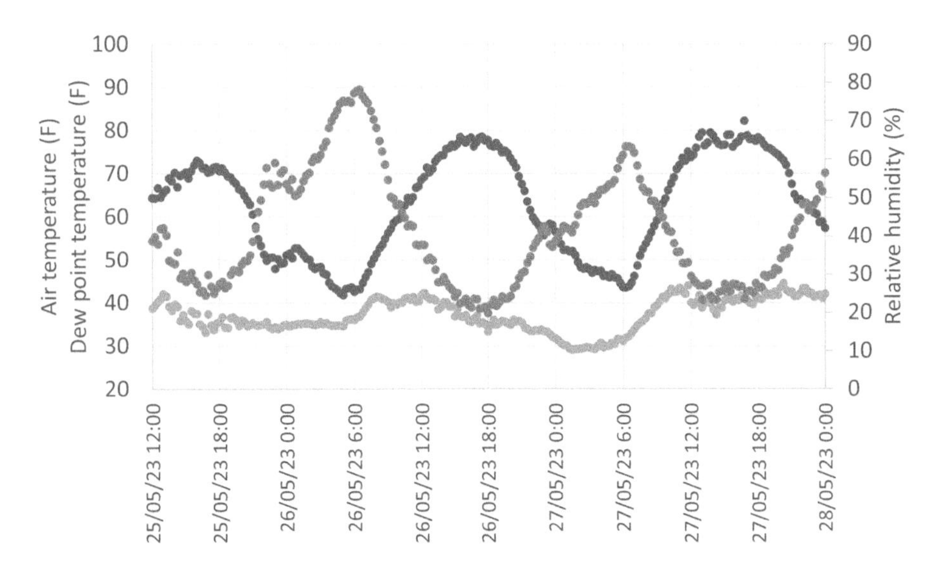

Figure 5.2 Typical daily variation in temperature and humidity. Air temperature (blue), relative humidity (orange) and dew point temperature (grey). The air temperature and relative humidity vary inversely when there is little change in water content of the air (represented by the dew point temperature).

the air to cool which will cause the RH to rise. If clouds form during the night (Figure 5.2 at time 0:00 on 23 May 2023), the loss of heat from the ground to the atmosphere will lessen and the air will warm and decrease the RH.

A single RH value (follow any colored line in Figure 5.1) can correspond to a wide range of moisture (vertical axis in Figure 5.1) and temperature conditions (horizontal axis in Figure 5.1). The other measures of humidity (dew point temperature, wet bulb temperature and specific humidity) do not change with change in air temperature and are therefore better representations of the actual water content of the air.

The moisture in the air over a location typically varies only slowly over time. If the amount of water in the air does not change during the day, the dew point temperature will not change greatly while the change in air temperature will cause a change in the RH because the amount of water the air can hold increases as the air temperature increases (Figure 5.1). Note that these changes in air temperature change the RH but do not change the dew point temperature (Figure 5.2).

If the air does not saturate at night, the variation in daily average RH will strongly depend on air temperature and not actual humidity in the air (Figure 5.3). However, if the air saturates, the dew point temperature will generally be the same as the air temperature. On a day-to-day basis, changes in water vapor (described by the dew point temperature) may or may not appear to change the RH. Assuming the **atmospheric pressure** does not change, the

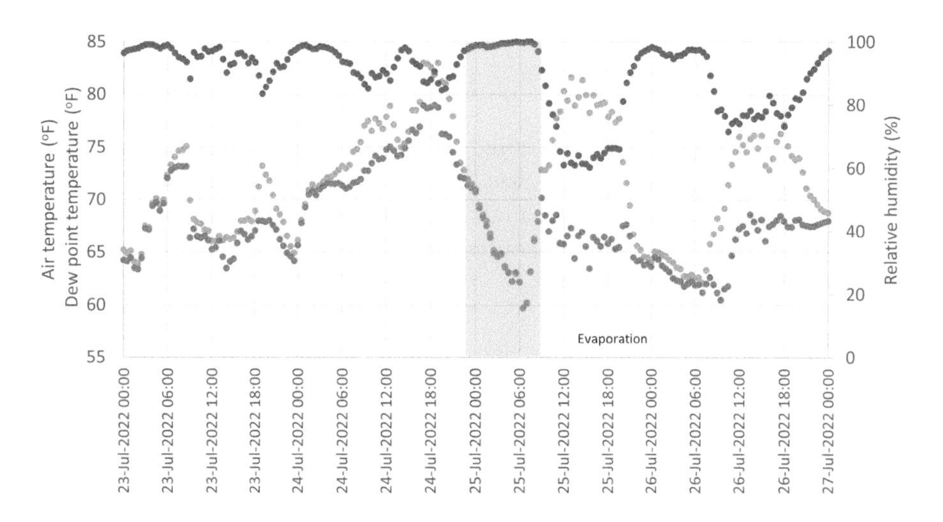

Figure 5.3 Daily variation in temperature and humidity under stormy skies. Air temperature (grey), relative humidity (blue) and dew point temperature (orange). Note the water vapor condensation (fog) after the rainy period (light blue region). Clearing skies warm the air and the fog is evaporated. The day-to-day water vapor content of the air (dashed line) is nearly constant (measured by the dew point temperature).

dew point temperature decreases by about 1.8 °**F** for every 5% RH at the Earth's surface.[3] At the Earth's surface, the wet bulb temperature decreases between 0.2 °F and 0.7 °F for every 5% of RH.[4]

Evaporation and condensation in the air

When the air saturates, water vapor is condensed out of the air into droplets. There is energy stored in the organization of the water molecules. This stored energy, called **latent heat**, increases or decreases as the water changes from vapor form to liquid form to solid form (phases of water). We cannot measure latent heat. However, when the water changes from vapor to liquid to solid, or *visa versa*, some of this latent heat is converted between sensible and latent heat (Table 5.1). So, changes in the phase of water in the environment result in changes in the sensible heat in the environment. Remembering from the last chapter, as sensible heat of any material changes, the temperature of the material changes depending on the **heat capacity** of the material.

Condensation transformations water from vapor to the liquid state. This occurs when the air cools to the dew point temperature (Figure 5.3). At this temperature the air becomes saturated (RH is 100%). Condensation in the air near the ground forms fog. This condensation adds sensible heat to the air (Table 5.1). Given the density of water droplets in dense fog,[5] the increased sensible heat in the air on condensation to fog increases the air temperature by about 0.2 °F given the heat capacity of air (Table 5.1).

Condensation as fog reduces the water vapor content of the air as indicated by the decrease in dew point temperature early on 25 July in Figure 5.3. That next morning the fog evaporates as the air temperatures rise with sunrise and the heating of the air by the sunlight. The reduction of sensible heat in the air due to **evaporation** however has little influence on the air temperature. Most moisture in the air is advected into an area: evaporation and **precipitation** (rain and snowfall) only 'recycles' 10% to 20% of water vapor in the local air.[6]

Table 5.1 Energy in phase changes of water

Process	Phase change	Sensible heat change per lb of water
Condensation	Vapor to liquid	+ 970 **BTU**
Freezing	Liquid to solid	+ 144 BTU
Deposition	Vapor to solid	+ 1053 BTU
Sublimation	Solid to vapor	– 1053 BTU
Melting	Soil to liquid	– 144 BTU
Evaporation	Liquid to vapor	– 970 BTU

Evaporation of the condensed water vapor in the air near the ground (fog) or liquid water on the ground increases the water vapor content of the air. The transition from water condensing on particles in the air to water evaporating from the droplet surface occurs when the air temperature is greater than the dew point temperature. This is evident during the morning of July 25th in Figure 5.3.

Evaporation also reduces the sensible heat in the air if droplets are evaporating from either the air itself or the sensible heat of the surface if evaporating from a wet surface (Table 5.1). Since there is less sensible heat, the air or surface temperature decreases when liquid water evaporates. But again, this loss of sensible heat results in only a small change in the air temperature- only about 0.2 °F decrease in the air temperature assuming evaporation of a dense fog.

Clouds are a result of water vapor condensation. As droplets in clouds grow from this condensation, they eventually may fall out of the cloud as rain. Rain also increases the water vapor content of the air as falling droplets evaporate (note increase in dew point temperature during the period of rain in Figure 5.3). This evaporation can cause fog.

Local humidity climates

We return now to the problem of estimating the RH where a body is found. An understanding of the influence of **local climates** on the RH is critical to having confidence in any calculated average value. Differences in the local humidity **climate**s of the measurement station and the location where the corps is found may be due to differences in elevation, differences in land use, proximity to water bodies, or differences in the terrain.

Due to elevation

Of the four measures of humidity described above, only specific humidity does not change with changing atmospheric pressure. To make humidity estimates for a corpse at a location at a different elevation from the measurement location, we must consider the effect of raising or lowering the air on the moisture content of the air.

Lowering the air warms the air at a rate of 0.55 °F per 100 feet in altitude (**adiabatic** warming) but does not change the water vapor content (or specific humidity) of the air, so the RH of the air decreases. Raising the air cools the air at a rate of 0.55 °F per 100 feet (adiabatic cooling) and may cool the air to water vapor saturation. If raising the air cools the air to its dew point temperature, the air will saturate and water vapor will condense out of the air into liquid form from that altitude on up. When water vapor condenses out of the air, some latent heat converts to sensible heat- heating the air it is condensing out of (Table 5.1). As a result, air that is rising and condensing cools at a rate lower than 0.55 °F per 100 feet: ranging from 0.2 to 0.3 °F per 100 feet. An approximation of the elevation at which condensation occurs is 225 times the difference between the air temperature and the dew point temperature near the ground surface.[3] The water vapor condensation is evident in the presence of clouds in the sky or fog on the ground if it occurs at a lower altitude than the elevation of the mountains or hills.

While the specific humidity of unsaturated air is unaffected by atmospheric pressure, the dew point temperature and wet bulb temperature are affected. Decreasing atmospheric pressure decreases the dew point temperature about 0.1 °F per 100 ft. Oddly, decreasing air pressure results in a lower wet point temperatures when air temperatures are less than about 70 °F and higher wet point temperatures when air temperatures above 70 °F.

Going back now to the situation of the corpse location and humidity measurement station location at different elevations, the humidity conditions at the corpse is determined after adjusting the station air temperature adiabatically and comparing the temperature with either the measured specific humidity of the air or the elevation-adjusted measured dew point temperature to the saturation specific humidity or air temperature of the location of the corpse.

- If the elevation-adjusted air temperature is less than the dew point temperature at that elevation or the saturation specific humidity at the elevation-adjusted air temperature is less than the measured specific humidity, then you can expect fog and a RH of about 100%.
- If the elevation-adjusted air temperature is greater than the dew point temperature at that elevation or the saturation specific humidity at the elevation-adjusted air temperature is greater than the measured specific humidity, the RH is calculated as a percentage of the measured specific humidity divided by the saturation specific humidity at the elevation-adjusted air temperature.

Due to clouds

Clouds are defined by water droplets or ice crystals that first appear above about 100 feet (as measured by ASOS- Chapter 7). The heights of the bottom of the cloud layers are **record**ed at airport ASOS stations up to an altitude of 12,000 ft.

If there are measures of the heights of the bottom (or base) of clouds as well as the humidity at a climate station, it is possible to estimate the likelihood of saturated water vapor at the altitude of the location of interest can be compared against the base of the clouds. Since clouds typically form as air rises on the windward side of mountains, measures of the **cloud cover** and **heights** made upwind of locations of interest in mountainous areas will likely not represent conditions within the mountains. Estimating clouds in mountainous areas where no cloud measurements are made should be based on cloud measurements made either within the mountainous area or on the leeward (downwind) side of the mountain.

The estimation of the RH at the ground in the mountains due to clouds depends in part on the spatial homogeneity. Measurements of hourly (or more frequent) cloud cover and height are made at airports (Chapter 7). The RH in liquid droplet clouds where the air temperatures are at least 23 °F is 99%.[7] Partial sky cover of liquid droplet clouds has an average RH of about 90%.[8] Clouds made up of ice crystals have air temperatures between −13 °F and −40 °F and average RH values greater than 100% relative to liquid water and RH of approximately 100% to 105% relative to ice.[7] Cloud in air temperatures of between 23 and −13 °F have frozen and liquid particles and have an average RH of 100%.[7]

Due to advection

We stated that **advection** is a sensible heat exchange process involving when air moving horizontally across the landscape. This air also has a water vapor content. So, water vapor is also advected by air movement. Between 80% and 90% of the water vapor in the air across most of the world is carried into an area by advection.[6] Changes in moisture advection are generally a result from the progression of large-scale weather systems across the landscape such as warm and cold **front**s. Figure 5.4 illustrates the advection of moisture (increasing in dew point temperature) from 30 July through 8 August 2022 due to winds from the Gulf of Mexico followed by the advection of dry air from Canada behind a cold front. Note that the advection of moisture (indicated by the increasing daily average dew point temperature) is reasonably well represented by the daily average RH.

The effect of a lake breeze on the humidity of the air depends on the character of the air overlying the land: the air over the land may have higher dew point temperatures than that over the water.[9] The RH commonly rises about 10% with the onset of a sea breeze.[10] A similar effect can be found as air is advected across well-watered or irrigated vegetation[11] and swampland. Evaporation and transpiration by vegetation can evaporate more water into the air than a flat-water body.[12] However, the change in humidity may not be detectable: a reservoir one-half mile across did not contribute enough water vapor into the air over two months to detect a change in specific or relative humidity on the surrounding land.[13]

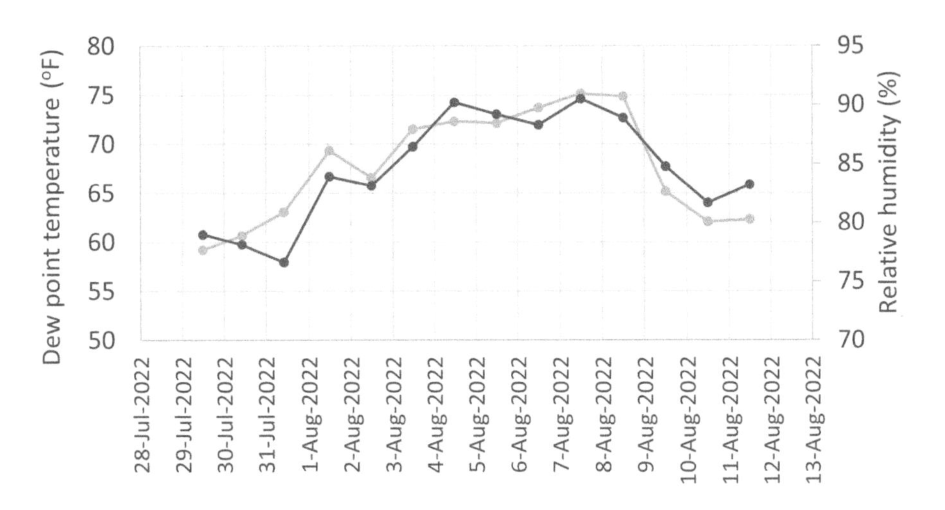

Figure 5.4 Influence of moisture advection ahead of a cold front in central Indiana. Relative humidity (blue circle) and dew point temperature (orange circle). Cold front passed through 9 August 2022.

Due to fog

An understanding of where and when fog might be expected helps assess the reasonableness of the humidity record. Fog may occur either due to adding water vapor to the air from advection or evaporation until the air saturates or cooling the air to the dew point temperature from advection or radiative cooling. Fog is more common in some local environments than others and is more common in some seasons than others. Various types of regional fog conditions and their seasonal occurrence across the US are discussed by Stone.[14] Here we briefly discuss the types of fog one can expect in **open terrain**, hills and mountains, coastlines, cities, as well as generally across the **landscape** due to changing weather:

Fog can form during the night when the radiative heat loss from the ground cools the air below the dew point temperature. This usually forms in open terrain (high **sky view**) under cloud-free skies. If the ground has some snow on it, the cooler surface of the snow compared to the bare ground will result in fog first forming over the snow. The average RH of this type of fog was 100% in a study in Virginia, USA.[15] This cause for fog formation represented 28% of fog events in the plains and hilly suburban and rural areas of Connecticut, New York, and New Jersey[16] and 78% of all fog events in Zagreb, Croatia.[17]

Another type of fog forms in low areas such as valleys and hollows. As the air cools it becomes denser. This density-driven cold air flows into the hollows and valleys and accumulates if there is no drainage outlet to a lower elevation.[18] These areas differ from the depressions where frost may form by their size- the larger width and depth of these areas allow for the accumulation of the cold air.

Fog can also form on the slopes in hilly and mountainous areas. As moist air is forced over a mountain or hill the air temperature drops adiabatically. If this is cooled below the dew point temperature, fog will form. This is most common during the winter and spring in the US. The RH of fog formed due to air being lifted over hills and mountains in Taiwan averaged 99%.[5] This type of fog is common in areas around New York City.[16]

Fog also can form as air descends. The air at the tops of the mountain or hill is colder than the slope because the greater sky view allows more radiational cooling at the surface. This cool surface air descends at night and can pool in valleys. As the cool air descends it mixes with the air lower down the slope and cools that air below the dew point temperature – and fog forms. This fog occurs most frequently in Croatia during in fall and winter.[17]

Fog is common near water bodies. Advection of nearly-saturated warm moist air over water bodies onto colder ground commonly forms fog as the cold ground cools the advected air to the dew point temperature. However, air coming off the ocean does not always form fog.[19] Fog occurs in summer and fall afternoons along the California coast: between 10% to 15% of time on the northern coast and 25% to 30% of the time along the southern coast.[20] This cause for fog formation is the most common in spring and fall in the coastal New York City area.[16] The RH of fog formed by advection off the Chesapeake Bay averaged 100%.[15]

Fog in urban areas is uncommon. While the surrounding rural landscape of an urban area is often vegetated with high water vapor content,[21] the RH of the air advected into the urban area is reduced as the urban area heats the air (increasing how much water vapor can be held before condensation). The RH in the urban area of Edmonton, Canada averaged 10% lower than the surrounding rural areas in the summer.[22] A study of fog formation showed it was infrequent in three cities in Germany.[23] Fog can however form in cities from other processes. The combined presence of coastal on-shore breezes and a density-driven cold air drainage result in fog formation in Perth, Australia.[24] Also fog in the urban center of Belgrade, Sebia[21] and across New York, Connecticut and New Jersey results from saturation due to the evaporation of rain droplets during storms.[16]

Air: ground moisture exchanges

Although estimates of the humidity available for decomposition are based on the humidity of the air, the humidity of a corpse lying on the ground will differ somewhat from that of the air over the ground. How does water vapor move from the air to the ground or ground to the air? Just as sensible heat moves down a temperature **gradient**, there is a movement of water vapor transfer of water vapor down water vapor gradients. The surface of liquid water corresponds to the saturation water vapor content for the temperature of the water surface. Since the saturation vapor pressure decreases with decreasing temperature, condensation on the ground surface (dew) will often occur before fog occurs on clear sky nights. Since fog limits the transmission of sunlight to the ground, the sunlight typically will evaporate the fog (often called 'fog lifting') before dew evaporates or snow melts or sublimes.

Dew

Dew will form as the ground surface temperature drops to the dew point temperature. This usually occurs in the late hours of the night before sunrise during the summer and fall seasons in the midwestern US (between hours 4 and 8 in Figure 5.5). Although the ground surface is at the dew point temperature of the air, the air temperature is often higher than the dew point temperature (Figure 5.5). Since it is the ground temperature, not the air temperature, that will initiate dew formation, dew is reported to form when the RH of the air is 91% to 99%.[12] The RH of the air, however, cannot be used as an estimate for the occurrence of dew. Almost all of the days with dew formation in a dry climate in eastern Oregon had RH above 90% but many days had RH above

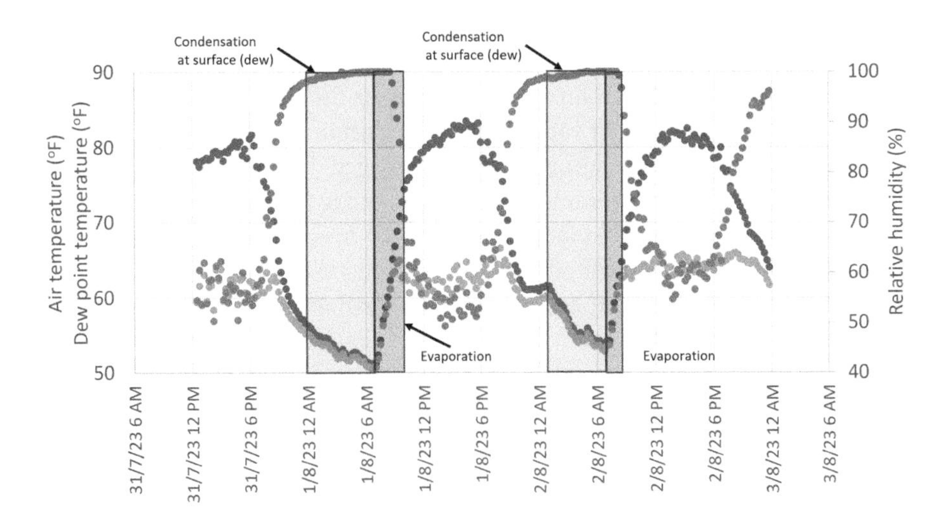

Figure 5.5 Daily variation in temperature and humidity with fog under clear skies. Air temperature (blue circle), relative humidity (orange circle) and dew point temperature (grey circle). Note the water vapor condensation (dew) each night and evaporation each morning. The amount of water in the air during the night decreases as vapor is condensed out of the air, but returns to the air in the morning as the dew is evaporated. The day-to-day water vapor content of the air (dashed line) is nearly constant (measured by the dew point temperature).

90% for more than four hours and without dew occurring.[25] Fog typically lasted between 10 and 17 hours in an area of coastal fog in California between April and October.[19]

As long as dew is depositing, the air temperature typically stays at or near the dew point temperature due to the contribution of sensible heat due to condensation offsetting the radiative cooling of the ground. As morning comes and the surface temperatures (sensible heat) increase due to heating from the sun- providing more energy to evaporate the liquid water on the surface into vapor. And since the air at the dew surface is saturated but the air over the surface is warming and hence becoming unsaturated, the liquid water evaporates into the air. The surface temperature usually also rises because the cooling of the surface due to the evaporation is less than the heating of the surface by the sunlight. Dew has been reported to last four to 14 hours in humid climates England[26] and up to 15 hours in semi-arid South Africa and Israel.[27, 28] Eight-two percent of the dew events over a 6-year period lasted more than 4 hours at a small town east of the Cascade Mountains.[25]

Dew droplets on a corpse may coalesce and pool in crevasses on the body. Consequently, condensed liquid water may be retained in the folds of the body longer than the dew on the ground. Because the sky view of the body surface influences the surface temperature (Chapter 4), the occurrence of dew on a body in urban areas, woodlands or forests will depend on its location on the ground.

Snow

When snow covers the ground, any object on that ground becomes insulated from changes in air temperature and humidity. The RH of the air *in* the snowpack depends in part on whether there are only frozen crystals or a mix of crystals and liquid water. The RH over a liquid surface within the snowpack would be 100% but that over crystal surfaces will be much more than 100% (110% at 14 °F). As a result, a corpse on the ground surface covered with snow might be expected to have a RH of at least 100%.

Melting of snow and ice does not directly change the air's water vapor content since it is changing from solid to liquid on the surface. Melting reduces the sensible heat in the water and increases the latent heat of the air (Table 5.1). Snowpack can melt at temperatures as low as 14 °F and stay frozen at temperatures as high as 40 °F,[29] depending on available sunlight and advected sensible heat. The addition of sensible heat to the ice and the transformation of the ice into liquid water will keep both ice and water at 32 °F until all the ice is melted. If a corpse is frozen, the melting of the frozen water in and around the corpse will therefore keep the surface temperature of the corpse at 32 °F until thawed even though the air temperature may be much higher than freezing.

Sublimation of ice or snow, transforming the solid water into vapor, occurs to some degree when surface temperatures are below freezing. Sublimation will increase as the temperature decreases and energy for the transformation is available. The source of energy is usually from sunlight and advected dry warm air.[30] Sublimation increases the humidity of the air and reduces the snowpack thickness – allowing the exposure of corpses to the air and sunlight and greater sensible heat exchange between the corpse and the air and soil. Sublimation also decreases the sensible heat and increases the latent heat (Table 5.1).

Moisture in the soil

The RH around a buried corpse depends on the surrounding soil's humidity. The pore space of an undisturbed soil varies with composition of the soil. Air-dried soils, found where plants have wilted, have RH values ranging from 20% to 55%.[31] Any visible amount of moisture in the soil results in pore space RH of 99% to 100%,[32, 33] so the RH of soils on which plants are not wilted is essentially 100%.[32]

When the ground freezes, soil pore spaces have a combination of ice, liquid water and water vapor.[34] Since the RH over ice is greater than 100% and the liquid water is super-cooled, staying liquid at sub-freezing temperatures, the RH in the frozen soil pore space exceeds 100%. In a study in Inner Mongolia, soil pore space water froze nightly as air temperature dropped below freezing and melted during the day when the air temperatures rose above freezing.[35] Soil water remained frozen during both night and day after approximately two weeks of daytime air temperatures below freezing.[35]

Impacts of mid-latitude large-scale weather systems on humidity averages

How does RH vary from day to day? It is common in mid-**latitude temperate climates** to have many days with little variation in daily average air temperature and RH punctuated by shorter periods of time where the large-scale weather systems change the daily average air RH and/or temperature. Since **PMI** determinations are based on average daily RH values, how should this be calculated? Since most measurements of humidity come from hourly records, the average daily humidity could be calculated either as the average of the hourly RH over the day or as the average of the minimum and maximum daily RH (as commonly done for daily average air temperatures). Comparisons between the daily average RH calculated from the hourly measured water vapor content and the average RH calculated from the daily maximum and minimum measured RH varied from a 20% overestimate to a 21% underestimate over a six-year period in southern Indiana. However, the rate of change of above-ground RH is influenced more strongly by the diurnal changes in air temperature than large scale weather changes. As a result, the **error** in estimating the daily average RH from the daily maximum and minimum RH and not the average of the hourly RH is relatively small. An assessment over 638 days at one location to the lee of the Rocky Mountain range (Appendix case 7) found an average 2% overestimate of the true daily average RH. An analysis of the difference in average RH for an 88-day summer and fall period in central Indiana found an average 5% underestimate by determining the average from the maximum and minimum daily RH values.

References

1. Chumlea, W.C., Guo, S.S., Zeller, C.M., Reo, N.V. and Siervogel, R.M., 1999. *Kidney International, 56*(1), pp.244–252.
2. Ciaffi, R., Feola, A., Perfetti, E., Manciocchi, S., Potenza, S. and Marella, G.L., 2018. *Romanian Journal of Legal Medicine, 26*(4), pp.403–411.
3. Lawrence, M.G., 2005. *Bulletin of the American Meteorological Society, 86*(2), pp.225–234.
4. Stull, R., 2011. *Journal of Applied Meteorology and Climatology, 50*(11), pp.2267–2269.
5. Gonser, S.G., Klemm, O., Griessbaum, F., Chang, S.C., Chu, H.S. and Hsia, Y.J., 2012. *Pure and Applied Geophysics, 169*, pp.821–833.
6. Trenberth, K.E., 1999. *Journal of Climate, 12*(5), pp.1368–1381.
7. Korolev, A. and Isaac, G.A., 2006. *Journal of the Atmospheric Sciences, 63*(11), pp.2865–2880.
8. Twohy, C.H., Coakley Jr, J.A. and Tahnk, W.R., 2009. *Journal of Geophysical Research: Atmospheres, 114*(D5).
9. Lyons, W.A. and Olsson, L.E., 1973. *Monthly Weather Review, 101*(5), pp.387–403.
10. Atkinson, B.W. 1981. *Meso-scale atmospheric circulations.* Academic Press, London, 495p.
11. Segal, M. and Arritt, R.W., 1992. *Bulletin of the American Meteorological Society, 73*(10), pp.1593–1604.
12. Rosenberg, N.J., Blad, B.L., Verma, S.R. 1983. *Microclimate: the biological environment.* 2nd Ed., John Wiley and Sons, New York, NY 495p.
13. Gregory, S and Smith, K. 1967. *Weather, 22*, pp.497–505.
14. Stone, R.G., 1936. *Geographical Review, 26*(1), pp.111–134.
15. Gerber, H.E., 1981. *Journal of atmospheric Sciences, 38*(2), pp.454–458.
16. Tardif, R. and Rasmussen, R.M., 2007. *Journal of Applied Meteorology and Climatology, 46*(8), pp.1141–1168.
17. Prtenjak, M.T., Klaić, M., Jeričević, A. and Cuxart, J., 2018. *Atmospheric Research, 214*, pp.213–227.
18. Cornford, C.E., 1938. *Quarterly Journal of the Royal Meteorological Society, 64*(277), pp.553–592.

19. Scherm, H. and Van Bruggen, A.H.C., 1993. *Agricultural and Forest Meteorology, 66*(3–4), pp.229–245.
20. O'Brien, T.A., Sloan, L.C., Chuang, P.Y., Faloona, I.C. and Johnstone, J.A., 2013. *Climate Dynamics, 40*, pp.2801–2812.
21. Vujović, D. and Todorović, N., 2018. *Theoretical and Applied Climatology, 131*, pp.889–898.
22. Hage, K.D., 1975. *Journal of Applied Meteorology and Climatology, 14*(7), pp.1277–1283.
23. Sachweh, M. and Koepke, P., 1995. *Geophysical Research Letters, 22*(9), pp.1073–1076.
24. Golding, B.W., 1993. *Monthly Weather Review, 121*(9), pp.2529–2541.
25. Crowe, M.J., Coakley, S.M. and Emge, R.G., 1978. *Journal of Applied Meteorology (1962–1982)*, pp.1482–1487.
26. Burrage, S.W., 1972. *Agricultural Meteorology, 10*, pp.3–12.
27. Lomas, J., 1965. *Agricultural Meteorology, 2*(5), pp.351–359.
28. Baier, W.J.A.M. 1966. *Agricultural Meteorology, 3*(1–2), pp.103–112.
29. Kuhn, M., 1987. *Journal of Glaciology, 33*(113), pp.24–26.
30. Strasser, U., Bernhardt, M., Weber, M., Liston, G.E. and Mauser, W., 2008. *The Cryosphere, 2*(1), pp.53–66.
31. Schneider, M. and Goss, K.U., 2012. *Geoderma, 170*, pp.64–69.
32. Campbell, G.S. and Norman, J.M. 1988. *An introduction to environmental biophysics*. 2nd Ed., Springer-Verlag, New York, NY. 286p.
33. Case, C.M. and Welch, A., 1979. *Journal of Hydrology, 43*(1–4), pp.99–120.
34. Yu, L., Zeng, Y., Wen, J. and Su, Z., 2018. *Journal of Geophysical Research: Atmospheres, 123*(14), pp.7393–7415.
35. Zheng, C., Chen, Y., Gao, W., Liang, X., Šimůnek, J. and Liu, X., 2023. *Journal of Hydrology*, 627, 130401.

Microclimate of the corpse location
What to consider

6

We have discussed how the local terrain influences the **local climate** air and soil **temperature** and **humidity**. Now we will consider some specific **climate**s resulting from the interaction of the corpse location and the air over the corpse- called **microclimates.** As discussed in Chapter 4, the energy budget of a surface includes the transfer of **radiant energy** from the sun and the surrounding environment, **advection** of heat by the wind, and buoyancy of the air, **conduction** of heat with the ground or other solid surface in contact with the body, and **evaporation** of any water from a surface. The moisture in the air depends on the evaporation as well as advection and **precipitation** discussed in Chapter 5.

Surface temperatures versus air temperatures

Since the earth (or a body) surface is the surface that receives **solar radiation** and emits **thermal radiation** (Figure 6.1), the temperature of a surface is typically warmer than the air above or ground below the surface during the day and less than the air above or ground below the surface during the night.

The air temperature over a corpse location differs from the corpse surface temperature in that the air temperature (Figure 6.1) represents the radiant heat exchanges with the body and the environment and the convective and buoyant heat exchanges with the environment (Figure 6.1). The daily average ground surface temperature averaged 3 °F (+/− 10 °F) above the air temperature over a wide range of vegetation types and complex terrain.[1] Within an oak forest, the air temperature differences between that at the surface were not significantly different.[2]

Convective transport of heat from the air to the corpse (Figure 6.1), partially driven by the winds, typically cools the surface during the day and heats the surface during the night resulting in average daily surface temperatures close to that of the air.[1] If the body is wet at the time of discovery or during the course of the air temperature and humidity assessment, evaporation (Figure 6.1) will reduce the temperature and increase the **RH** during the day.[3]

Portions of a corpse surface may be warmer due to dense maggot masses. *Calliphoridae* and *Lucilia* maggot mass temperatures on pig carcasses rose to between 9 °F and 34 °F above the air temperature[4–8] while masses of

DOI: 10.4324/9781003486633-6

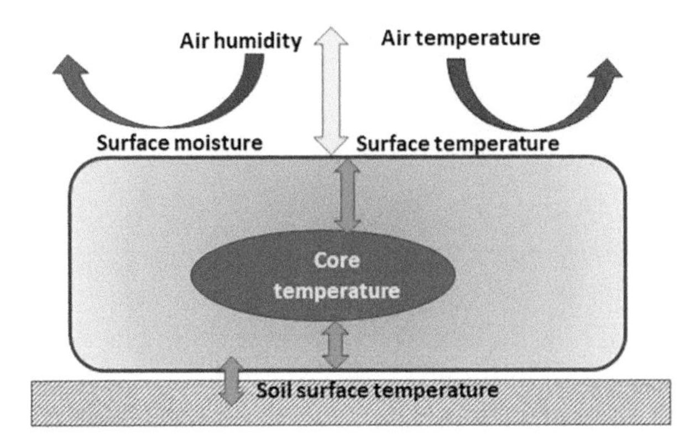

Figure 6.1 Relationship of energy exchanges and the temperature of a 'stylized' corpse with the ground surface and air temperature. Conduction (green arrows), convection (orange arrow), evaporation (blue arrow) and **net radiation** (yellow arrow) heat exchange are indicated.

Calliphoridae and *Chrysoma* have been reported to be 20 °F above the air temperature.[8] The surface temperature of a *Calliphoridae* maggot mass appeared to peak in air temperatures around 75 °F.[4]

Maggot mass temperatures however do not strongly influence the overall corpse temperature as they typically cover a relatively small fraction of the overall corpse area.[9, 10] While maggot masses on pigs showed elevated temperatures of 14 °F to 18 °F from the surrounding soil surface temperature, the entire pig carcass was elevated only 2 °F to 6 °F.[11] Maggot mass temperatures on sunlit carcasses have been observed to be 3 °F to 6 °F higher than on shaded carcasses.[4] The maggot mass temperatures were also found to be influenced by **cloud cover.**[12]

The thermal mass of the corpse may influence the heat exchanges between the surface and the core (Figure 6.1). Temperature **gradient**s between the core and surface of sheep carcasses suggest that the surface heated the core during the day and the core heated the surface during the night.[13]

Microclimates in homogeneous terrain

Most bodies in serial homicides discovered after 7 days are disposed of outside.[14] Characteristics of the microclimates outside follow.

At the ground in open terrain

Some corpses are discovered in isolated areas that are relatively free of vegetation. These areas are considered to be 'open' terrain. The temperature of bodies discovered in the open will be primarily be a result of the radiant heat exchanges from the sun and the environment, **convection** of heat with the wind, and conduction of heat with the ground. Solar radiation, the dominant incoming **radiation** in **open terrain**, and is strongly influenced by slope and aspect of the terrain.[15] Average daily air temperatures on the north and south slopes of an east-west valley varied by 5 °F.[16] While the humidity of the air will not be influenced by slope and aspect, the RH will be influenced by any change in the daily air temperature.

When the corpse is lying on the bare ground, the conduction of heat to and from the ground can be significant. This will likely result in the average temperature of the corpse and the ground to be very similar. If the corpse is lying in dead leaves, branches, or crop residue, the material will tend to insulate the body from heat exchange with the ground (Table 4.1), resulting in the corpse being close to the air temperature.

Sky view is important in the exchange of radiant energy (Chapter 4). Low grass areas (where upper surface of the corpse is greater than half the height of the grass canopy top) are similar to most climate measurement stations. Consequently, the air temperature and RH at such locations can be directly estimated from the measurement location. If the corpse is lying in high grass (where upper surface of the corpse is less than half the height of the grass canopy top), then the microclimate of the body should be treated like vegetated land discussed below. If the corpse is discovered in dry desert or barren land (free of vegetation) such as a construction site, the radiant heat load on the barren landscape will increase the air temperatures and correspondingly decrease the RH.

To estimate the microclimate of a corpse found in the open in a future case, we need a description of the local environment including:

1. The wetness of the ground around the body
2. The sky view over the body
3. The type of vegetation (deciduous, coniferous, crop) blocking the view of the sky
4. Characterization of the soil type (sandy, organic, silty)
5. If sloped, slope and azimuthal orientation of the terrain

At the ground in vegetated terrain

Corpses are often left in rural vegetated areas.[17, 18] The energy budget of the air under a vegetation canopy is predominantly driven by convective, conductive and thermal radiative exchanges. To estimate the temperature and humidity microclimate of a body found in a vegetated area, we must first consider the height and distribution of the vegetation canopy (Figure 6.2) and the thermal radiation exchanges with the forest and exposed sky during the day and night.

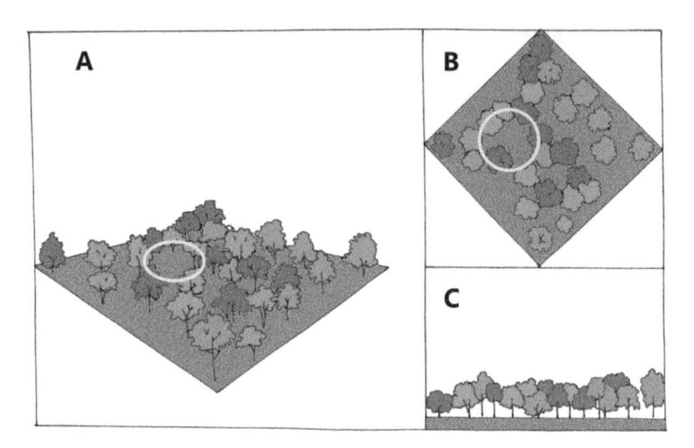

Figure 6.2 Canopy structure of forests. The quarter view (panel A), plan view (panel B) and profile (panel C) of the canopy are indicated. Canopy gaps are evident in panel A and B.

Illustration by E. F. Grant.

If a corpse is found in cropland or grasses where the top of the body is less than one-half the canopy height, we also must consider temporal changes in the overlying vegetation. Canopy density vary by crop and crop growing season.[19, 20] Care must be made to document the crop growing there and determine the density of the crop vegetation over the body at the time of discovery. If the date of the victim's last sighting is early in the growing season the date of planting of the crop is also needed.

Convective heat exchange with a corpse in the forest is limited since winds at the forest floor are generally light: less than 1 mph in many forests even when winds in the open were much higher.[21] Canopy gaps (Figure 6.2) also generally have low wind speeds: gaps up to 3078 ft^2 had only 1 mph wind speeds.[22]

Actively growing vegetation during the summer[23] and greater retained snow cover during the winter[24] will often reduce air temperatures in the open compared to the forest. In general, the air under forest canopies is 4 °F to 5 °F warmer in winter and 2.0 to 4.6 °F cooler in summer than in the open.[2, 15, 24, 25]

While sky view strongly influences radiative heat exchanges in the open, the density and gaps of a vegetation canopy (Figure 6.2) do not tend to affect air temperatures under the canopy significantly. Gaps in the in-leaf deciduous forest had average air temperatures within 0.9 °F of the surrounding forest (Abd Latif and Blackburn, 2010). The 30-year average daily temperatures under three coniferous forest management schemes (partially harvested, **understory** burned, and dense mature) were within 1.3 °F of the average in a clearing.[26]

Humidity in a forest is relatively insensitive to the forest density with **RH** on the forest floor ranging from 90% to 98% over spring, summer and fall.[27] **Relative humidity** under discrete trees or clumps of trees in a suburban setting average 81% while air in the open averaged 59% during the summer in a temperate mid-**latitude** location.[28] During the day the crop vegetation will increase the humidity of the crop canopy with the highest humidity near the ground as in the forest during spring and summer.

To estimate the microclimate of a corpse found on the forest or cropland ground in a future case, we need a description of the local environment including:

1. The height and density of vegetation above the body
2. If cropped land, type and stage of growth of crop
3. If forest, type of forest (deciduous or coniferous)
4. The wetness of the ground around the body
5. The material (bare ground, dead plants…) the body is lying on

Under snowpack in open or vegetated terrain

If a corpse is covered by snow, the snow will insulate the corpse from changes in air temperature and reduce soil temperature variability under the body. In one study, the 0.4-inch depth soil temperature under a 3-inch snowpack varied only 28% of the air temperature variation while a 5-inch snowpack essentially decoupled the air and soil temperatures with the soil temperature variation only 3.6 °F or 10% of the air temperature.[29]

Under snow cover the temperature will likely be at or below freezing until air temperature exceeds around 39 °F.[30] The presence of heat production by larval masses will also allow development of larvae at temperatures around 39 °F.[31] The RH of air pockets under snow depends on the nature of the water surfaces: 100% if the particle surfaces are liquid and greater than 100% (and dependent on temperature) if the surfaces are solid. A corpse, however, will thaw at temperatures lower than 32 °F due to the organic compounds in the body liquid solution.[32]

To estimate the microclimate of a corpse on the ground surface but under snowpack in a future case, we need a description of the environment including:

1. the sky view over the snowpack
2. The type of vegetation (deciduous or coniferous) blocking the view of the sky
3. the depth of the burial relative to the surface of the snow

Under the ground in open or vegetated terrain

A buried corpse exchanges heat with the surrounding soil by conduction. The average daily soil temperature is influenced by heat stored in the soil and **weather** conditions such as changes in air temperatures, cloud cover, rain, and snow cover.[33] Soils with small sky views are typically close to the air temperature, influencing soil temperatures to at least 10-ft depth over the course of the year.[34]

Table 6.1 Terrain surface classifications

	Smooth	Rough	Moist	Dry	Hot	Cool
Bare soil	×			×	summer	Winter
Parking lots	No vehicles	Vehicles		×	×	
Roads	×			×	×	
Crops		×	green		brown	green
Forest/ woodland		×	green		Brown	green
Shrubs		×	green		Brown	green
Grass	cut	uncut	green		Brown	green

The annual variation of daily soil temperatures lag that of the air as a result of the higher **heat capacity** of the soil compared to air (Table 4.1). In **temperate climates**, the soil temperatures are generally higher than the average daily air temperature during the winter and spring and lower during summer and fall.[34] Annual average 8-inch depth soil temperatures are higher than air temperatures in both cold and hot climates: 9 °F higher in cold and 5 to 7 °F higher in hot dry climates.[35] Across the US, the average annual soil temperatures at 8-inch depth are 2 °F higher than the air temperature.[35]

Since the **conductivity** of the corpse skin is comparable to that of soil (Table 4.1), the average surface temperature of the body will be near that of the surrounding soil. However, corpses may have elevated core temperatures over the average soil temperature: over nine weeks corpses averaged 15 °F, 9 °F, and 6 °F above the soil temperatures at 1 ft, 2ft, and 4 ft depth respectively.[36] This was attributed to decomposition with an added influence of insect activity for the one-ft depth burial.[36]

Rainfall and meltwater contribute to soil moisture. If the soil is saturated with liquid water, the **soil pore** RH is nearly 100%. If the soil is not saturated and plants are not wilted, the humidity of a soil will be close to 99%.[37] If a corpse is buried, the soil above the body will often be less compacted and likely have lower RH compared to an undisturbed soil if not saturated.[38]

If the ground where the corpse is buried is frozen, the RH of the corpse surface will be at least 100% (see Chapter 5). An assessment of the soil temperature variation is especially important where the soil freezes during part of the year. Freeze/thaw cycles of the soil surrounding the corpse also contributes to the rate of decomposition.[39]

To estimate the microclimate of a buried corpse, we need a description of the environment including:

1. the sky view over the soil
2. the depth of the burial
3. the apparent compaction or looseness of the soil over the burial
4. the apparent dryness of the soil

In water bodies in open or vegetated terrain

Many bodies in serial homicides discovered after 7 days are disposed of in water bodies.[14] One quarter of homicide victims in rural Finland are found in water bodies.[17]

All water bodies exhibit temperature variations over the course of a day and year.[15] Freshwater bodies can be classified as those that are 'still', such as lakes and ponds, and those in motion such as streams and rivers. Daily variation in temperatures is typically greater for moving water than still water. Water temperatures of moving water depend largely on the **net radiation** along the stream and the advection of heat from upstream[40] while that of still water is determined largely by net radiation and heat storage.[41, 42]

Water temperatures of rivers or streams often depend on upstream tributary conditions including weather conditions and terrain (Figure 6.3).[43, 44] Shading of the stream typically reduces the temperature downstream.[45] Water moving from a 100% shaded stretch to a sunny stretch can change the temperature 30 °F per mile of river.[43] Shallow streams and rivers are also affected by solar radiation heating the streambed which then heats the water.[45, 46] Deeper or opaque water rivers and streams are generally not heated by the river- or streambed. Water temperatures of rivers and streams vary laterally as the depth of the water changes. For example, the Danube River surface water temperature can vary by 29 °F across it's width.[46]

Water temperatures in the headwaters of many streams are close to the groundwater temperature.[47] Groundwater temperatures vary across the USA from 40 °F in **cold climates** to 75 °F in hot climates.[48] As water moves downstream, its temperature typically increases at 1.8 °F per mile for small streams to 0.3 °F per mile for large rivers.[47] Stream and rivers lag the daily air temperatures by 0 to 8 days.[49, 50]

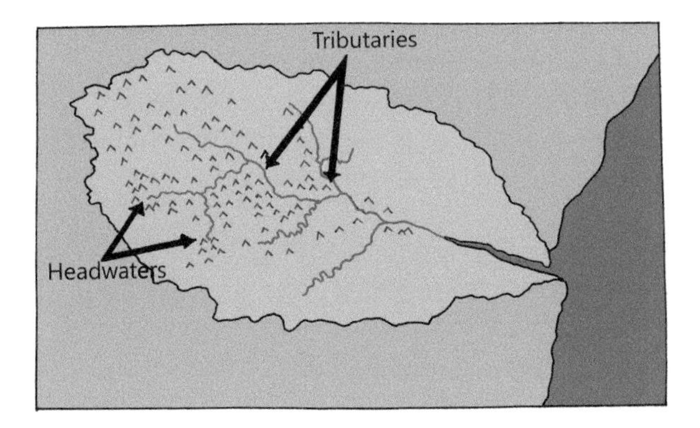

Figure 6.3 Example of river tributaries forming a watershed.

Illustration by E. F. Grant.

The daily range in water temperature is strongly related to the size of lakes and ponds. A small 16-ft deep 2.5-acre agricultural reservoir in Spain had differences in daily air versus water temperatures of up to 5.4 °F.[42] The daily range in summer surface water temperatures of 100 lakes ranged from 13 °F for ponds less than an acre to 2 °F to 4 °F for lakes 24 acres or more.[51] Vegetation near the shore of large water bodies or in shallow water bodies also tend to be warmer than open water: from 2 °F to 4 °F, up to 11 °F if the water is calm and the vegetation is dense.[15]

While the temperature of agricultural reservoirs and lagoons are generally sealed from **infiltration** of underlying groundwater, shallow lakes with depths ranging from 5 ft to 17 ft and areas ranging from 17 acres to 88 acres have been found to be substantially influenced by groundwater temperatures.[52] Water temperatures commonly lag the air temperatures annually with average monthly water temperatures exceeding the air temperature in temperate climates during late summer and fall.[53] A 154-ft deep lake in Russia has a lag in water temperatures of two weeks compared to air temperatures.[54] Average monthly surface water temperatures of deeper water bodies vary similarly to air temperatures: changing 1.1 °F to 1.4 °F for every 1°F air temperature change for lakes 84 to 175 feet deep.[53]

Corpses found in water bodies, either whole or in part, will largely be at the temperature of the surrounding water.[55] The dominant heat exchange between the environment and the corpse is conduction and convection with water with the addition of radiation and convection with the air if exposed. Heat convective exchange with moving water is about 100 times greater than still water.[56, 57]

To estimate the microclimate of a body in water, we need a description of the environment where the body was found:

1. exposure to the sky of the corpse
2. presence of buildings, forest, woods around the corpse
3. exposure of the body to the air
4. the depth and breadth of the river/creek at the time of discovery
5. the speed of flow of the river/creek at the time of discovery
6. the visibility of the bottom of the water body

Microclimates due to transitions in terrain/land use

Accessibility to a place to dispose of a body is also a factor in where corpses are found. For instance, a corpse may be just inside a vegetated area or along the shore of a creek.[18] The energy budget of the air over bodies found near changes in terrain (Table 6.1) may be strongly influenced by the upwind terrain. The air temperature and humidity of air flowing from one land use to another is related to the type and magnitude of change across the transition: smooth to rough or rough to smooth, moist to dry or dry to moist and hot to cool or cool to hot (Table 6.1).

The influence of one land use type on the temperature and humidity of air over another type largely depends on the difference in microclimates and the strength and frequency of the wind blowing air from one type to another across the transition. The influence of the upwind environment decreases as you go away further from the transition (Figure 6.4). A generally-accepted rule of thumb is that the influence of upwind environment on the air temperature and humidity will not influence the air over the new downwind land use after a distance of at least 200 times the change in roughness height of the two surfaces.[3, 58] For example, the air temperature influence of upwind cropland to downwind arid dry land was found to be 160 to 200 ft downwind of the transition.[59]

Since the corpse is on the ground, the influence of an upwind terrain change is likely to be evident only within about 300 ft of the change. Since air temperature and humidity measurements are commonly measured between 4 ft and 7 ft **agl**, the distance from the terrain transition (from surface 1 to surface 2, Figure 6.4) to the measurement location needs to be at least 800 ft to 1400 ft to represent the air over the corpse. If it is measured closer to the transition, the measurement will describe a mix of the upwind conditions (over surface 1 in Figure 6.4) and the downwind conditions (over surface 2 in Figure 6.4).

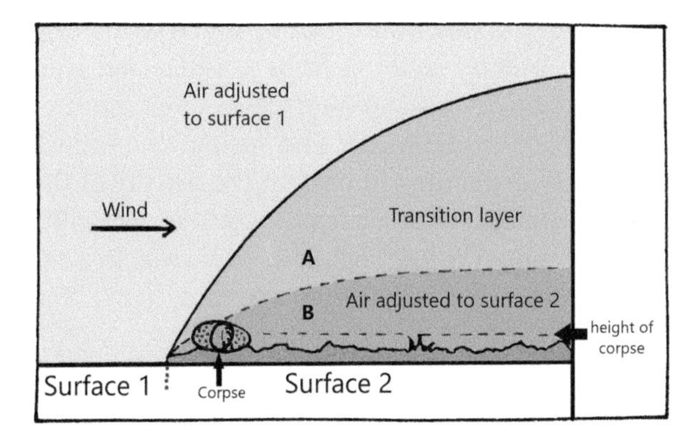

Figure 6.4 Example of advection of air across a transition in terrain. The air where the corpse lies in the air adjusted layer to the new surface 2 may differ from that measured at heights 'A' or 'B' on the surface 2. Measurements at height A do not represent the humidity and temperature conditions where the body is located while height B does.

Illustration by E. F. Grant.

At the ground near vegetation-open transitions

The advection of heat and moisture from open land into vegetated land downwind depends somewhat on the extent of the upwind terrain and the magnitude of difference in temperature and/or humidity between the two (Table 6.1). If the downwind vegetation has open space under the canopy, the wind will penetrate up to six times the height of the downwind vegetation.[60–63] In a 5-year study showed air temperature were less than 0.4 °F different than that within 6 times the height of the conifer forest[60] regardless of wind direction into or out of the forest. Air temperatures across a forest edge were very similar to that in the open during the winter but about 2 °F cooler than the open during the summer when at least 200 ft into the forest.[64] The air temperature less than one tree-height into a deciduous woods averaged 0.6 °F +/- 5.4 °F below the open air temperature over the winter.[64]

The RH across the edge of a conifer forest is also very similar to that in the open during the winter but about 9% higher least 200 ft into the forest.[62]

To estimate the air temperature and humidity for a corpse found in this microclimate, we need baseline environment information on:

1. Distance from the forest/woodland/cropland edge to open terrain
2. Density and height of the vegetation canopy and presence of an open understory
3. Type of forest (Coniferous or deciduous) or crop

At the ground near water-land transitions

Since the conditions over a water body can be much different from that over land, a lake or ocean breeze resulting from a water-land transition can be significant. The advection of heat and moisture is more significant during the summer and fall and less in the winter and spring. The influence of a small stream can vary widely depending in part on the density of the surrounding vegetation: a stream influenced the air temperature and RH 500 ft away in the open understory of a conifer forest[62] but only 25 ft away in a riparian forest.[65] In a forest gap, the influence of the stream was less than 2 °F within 50 ft of the stream.[66]

The influence of the water body on the nearby land also depends on how well the air over the water 'adapted' or 'adjusted' to the water as it flows over the water body. A distinct lake breeze is detectable only when the width of the water body is at least one mile across and extends inland by less than one-half the diameter of the water body.[67] The RH around a reservoir only one-half mile across did not show influence of the water body while the daily range in air temperatures was reduced from ranging 10.5 °F on the upwind side to 6 °F on the downwind side.[59] The influence of the lake breeze from the Great Lakes (with widths from 53 mi to 183 mi) extends approximately 50 mi inland with the daily maximum temperatures on land decreased by up to 4 °F upwind and downwind and the daily minimum air temperatures increased 2 °F both upwind and downwind.[68] Ocean breezes extended 12 mi to 30 mi inland in the midlatitudes.[69]

To estimate the air temperature and humidity for a corpse found in this microclimate, we need baseline environment information on:

1. Distance to the water body
2. Width of the water body
3. Type of water body (river, lake...)
4. Any information on the water body temperature.

Microclimates in man-made structures

Energy budgets in structures are driven largely by the radiative exchange with the walls and ceiling, the conduction of heat with the floor (if the body is lying on the floor) and the **infiltration** of heat through any open passages with environments differing from that where the body is found. **Condensation** and evaporation on the surfaces in the enclosure may also occur.

In vehicles

Corpses are occasionally found in vehicles.[70, 71] The temperature of bodies stored in vehicles, either in the trunk or in the cabin depend on the location of the body, the color of the vehicle, the tinting of the glass, and any gaps in windows that allow air to move into and out of the vehicle. The rubber tires act to insulate the car from the surface the car is sitting on. The energy budget for vehicles is strongly influenced by solar radiation and thermal radiation through the glass and heat conducted through the metal body (Figure 6.5).

The color of the vehicle strongly influences the air temperature in the cabin because of the influence on radiation **absorptivity**. A black car cabin averaged 9 °F above that of a white car during the summer,[72] and 13 °F in the spring.[15] At noon, a white car cabin with windows closed averaged 22 °F higher than the air outside while a black car cabin was 35 °F higher than the air outside.[15] A dark blue car cabin in the summer had maximum daily temperatures 38 °F higher in the sun than the shade while minimum daily temperatures were 2 °F lower in the sun than one in the shade.[73] The heat stored in a car appeared to result in a lag in a rigorously-modeled cabin temperature of 2 hours[72] – effectively not influencing the daily average temperature estimations for the car cabin based on outside environmental conditions.

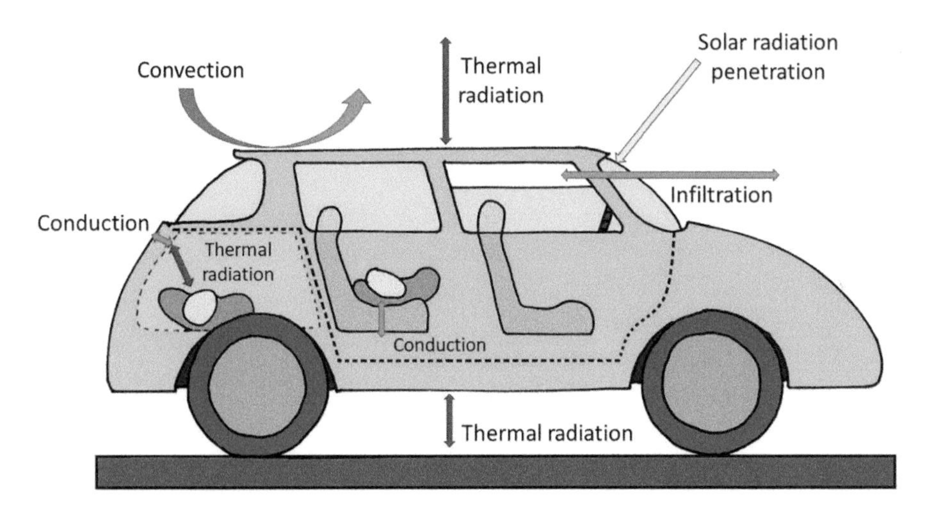

Figure 6.5 Heat exchanges outside and inside of a car for the estimation of a corpse (pink) located either in the passenger compartment or trunk. The major energy exchanges are indicated.

Illustration by E. F. Grant.

Partially shaded cars still receive substantial heating from solar radiation. Minimum daily temperatures in the cabin of a white car in bushes in the fall were within 2 °F of a nearby weather station while maximum daily temperatures were approximately 45 °F higher.[71]

Solar radiation through car windows have been reported to add about 5 °F to the cabin air temperature on the sunny side of two cars.[15] As a result, the orientation of the car is important in estimating air temperatures: cabin air temperatures changed by 6 °F after rotating a white car 90 ° in azimuth and by 2 °F after rotating a blue car 180°.[74]

If windows or doors are open, then additional heat is advected either into or out of the vehicle through infiltration (Figure 6.5). Since cabin air temperatures generally exceeds that of the outside, opening the windows will cool the cabin. Opening a window only one inch decreased the cabin temperature of a sunlit car 5.4 °F.[72]

If the corpse is in a car trunk, the body will come into a radiative equilibrium with the sides, top and bottom of the trunk (Figure 6.5). The temperature of the inside surfaces of the trunk will be essentially that outside due to the high conductivity of steel (361 **BTU** in/ (hr ft^2 °F); [75]). The thermal radiation from the metal surfaces dominates the energy budget. Assuming similarity of a gas tank and trunk of a car, the temperature in the trunk was within 3 °F of the outside air temperatures around a shaded car parked in an asphalt lot in the summer.[73]

To estimate the microclimate of the corpse in a vehicle, we need baseline environment information including:

1. Color of the vehicle
2. Location of corpse in vehicle
3. Any paths for air to move from where the body was found to the outside (open windows, broken glass, open doors, etc.).
4. The time period in which the vehicle was parked or in motion between the date of last sighting and date of discovery.
5. Shading of the vehicle by vegetation or buildings

In enclosed containers

Corpses are occasionally hidden in uninsulated shipping containers, industrial drums, garbage cans, garbage dumpsters or suitcases.[76] The energy budget for containers is strongly influenced by radiant heating by the sun and the environment with some heat convection associated with local winds. The temperature inside the container is a result of the conduction of heat through the walls and a thermal radiation from and to the surrounding environment.

Internal temperatures of an intermodal shipping container indicated a strong influence of solar radiation during the day with slow cooling during the night.[77] The daily range in temperatures was 60 °F and the maximum temperature gradient between the inside and outside of the container of 50 °F.[77]

High-density polyethylene industrial drums exposed to the direct sunlight showed wide variation in temperatures with average inside temperatures 2 to 4 °F +/− 29 °F warmer than the surrounding air.[78] Air temperatures inside a shaded trash can exposed to very light winds were 0.4 °F +/− 9 °F warmer than outside.[64] The interior temperatures lag the outside air temperature by a day or so[64, 78] with similar time lags in the increase or decrease in the air temperatures inside the trash can and the industrial drum. Modeling showed that convective cooling by winds had little impact on the inside temperature of high-density polyethylene industrial drums until winds exceed 12 mph.[78] Similar results might be expected for hard-shell suitcases.

Soft-shell suitcases exposed to the direct sunlight (when present) over 100 days were 14 °F +/- 18 °F warmer than the surrounding air.[78] The air temperature in five shaded soft-shell suitcases containing pig carcasses were not significantly different than the outside air.[79] The temperature inside a soft-shell suitcase averaged 8.8 °F above the surrounding air on a clear day and 7.4 °F above the surroundings on a partly cloudy day.[80]

The RH in sealed containers like drums and garbage cans will be driven by the moisture from the corpse and likely 100%. If the container is not sealed, the outside air will also influence the humidity. The RH in 5 soft-shell suitcases with pig carcasses averaged 81% after 14 days increasing to 90% after 42 days while the RH of the environment averaged 61% to 72%.[79]

To estimate the microclimate of a corpse within a container, we need to know baseline information including:

1. the composition of the walls of the container
2. the exposure of the sides and top of the container to the sunlight
3. the tightness of the seal on the container

In buildings and building spaces

Over 20% of bodies in serial homicides discovered after 7 days are disposed of inside buildings.[14] A corpse found in a building will be influenced by the buildings surroundings and construction, the type of building (temperatures controlled or not), the ventilation or infiltration rate, and the location of the body in the building.

The energy budget of the outer walls and roof of buildings include the influence of solar radiation, convection and advection (Figure 6.6). A building usually radiates more heat to the environment (thermal radiation, Figure 6.6) than it receives from the environment.[3] Consequently, the daytime solar radiation offsets the radiant heat loss from the building and is a major heat

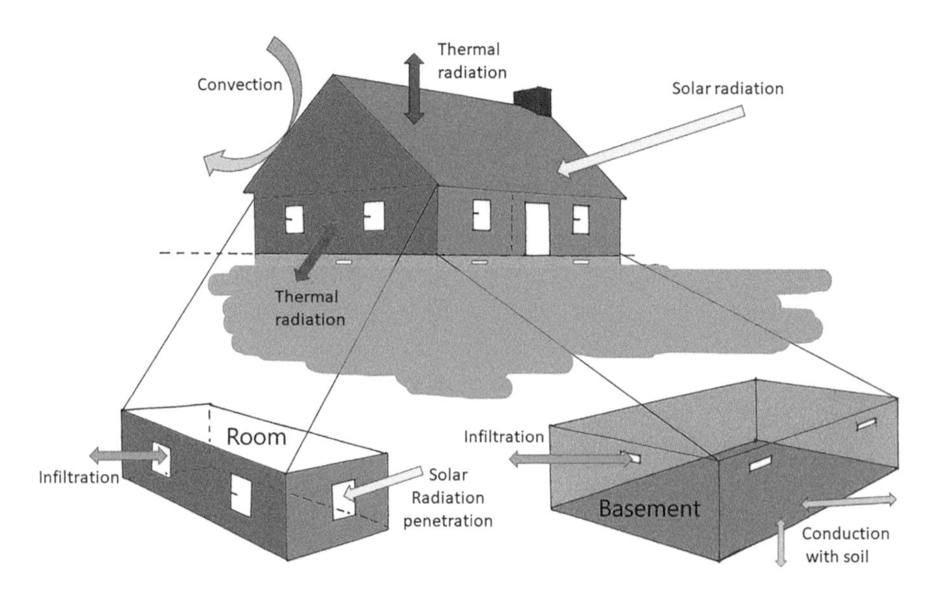

Figure 6.6 Heat exchanges for the exterior surface, an interior room, and the basement of a house for the estimation of the air temperature where the corpse might be located.

Illustration by E. F. Grant.

gain for the building exterior. At night the building exterior is usually losing heat to the environment. The type and size and position of windows is important because solar radiation penetrates into the building through them (Figure 6.6). Surrounding trees and/or buildings reduced the received solar radiation on the building exterior and reduced the air temperature variation in the building.[81] Reduced solar radiation loading of multi-family structures due to shared walls have narrower ranges in indoor temperatures than single-family dwellings.[82]

The air temperature within the building is also influenced by the conduction of heat through the floor, outer walls and roof. The air temperature and humidity in a building is influenced by infiltration of outside air into the building through windows and doors (Figure 6.7) and any heating and air conditioning.

Basements and Cellars

Buildings may have rooms below the soil grade (basements or cellars) (Figure 6.6). The air temperature and humidity in these spaces depend partly on the conduction of heat from the surrounding soil through the

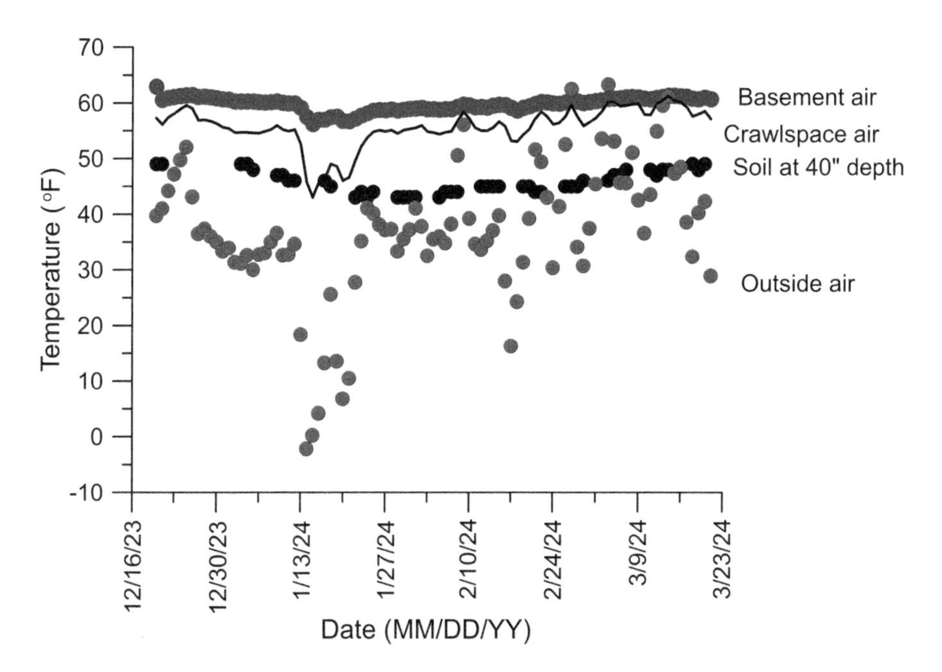

Figure 6.7 Winter temperatures in a Midwestern house crawlspace ventilated into the conditioned basement and an unfinished conditioned basement relative to the outside temperatures.

Data source for air temperatures: author R.H Grant. Data source for soil temperatures: Indiana State Climate Office.

walls and floors (Figure 6.6). Basements can be classified as **unfinished** and **conditioned**, **finished** and conditioned. Cellars are **Unconditioned** and unfinished spaces.

Cellars have ceiling, walls and floors of rock or soil and are often carved out of hills. Cellar air temperatures are generally very stable but the access tunnel allows significant air infiltration making their average temperature similar to the outside air temperature. Three cellars at a depth of 10.5 ft in a hot dry climate varied in air temperature over the course of a day, week, and month by 0.7 °F, 1.8 °F, and 4.7 °F respectively with average temperatures 1 °F cooler than average outside air temperature.[83]

Basement air temperatures are only weakly influenced by the outside air. The air temperature of an unheated unfinished residential basement changed 1 °F for every 5 °F change in the outside air.[84] An unfinished basement in a conditioned house in the midwestern US increased in temperature 1 °F for every 2.2 °F increase during the summer and decreased 1 °F for every 14.2 °F decrease during the winter.[64] Over long periods there was only a 5 °F difference in air temperatures outside compared to inside an unfinished unconditioned basement.[85]

When a concrete wall and floor in an unfinished basement is not ventilated, the air temperature will be similar to the soil temperature. Basement walls heat the basement air to temperatures above that of the outside air during the winter.[86] However, basement air temperatures change slowly due to the soil surrounding it (Figure 6.7). A conditioned unfinished basement cooled during the winter only 0.4 °F for every 1 °F decrease in 40-inch depth soil temperature and warmed during the summer only 0.5 °F for every 1 °F increase in soil temperature.[64] Another study showed the air 1.6 ft below grade in one basement was 5.6 °F cooler than the outside air while the air at 5 ft below grade was 0.7 °F cooler.[87] However, a finished basement can have significant heat loss to the soil in the winter and spring.[88]

Finishing the basement (insulating the walls and floor) essentially eliminates the air temperature gradients with height in the basement[85] and reduces the air temperature variation. A finished basement in a well-insulated and air-conditioned house in a cold temperate climate varied +/− 2 °F as the outside air temperatures varied from −13 °F to 94 °F.[89]

The assumed RH of unconditioned residential basements is 99%.[90] Finished basements in temperate climates typically control the RH by air conditioning in the summer and humidifiers in the winter. A finished basement in a well-insulated air-conditioned house in a temperate climate varied from 75% to 12% RH while the outside RH varied from 30% to 100% while the specific humidity outside was greater than the basement in the winter and less than the basement in the summer.[89]

A corpse in a cellar will largely assume a temperature corresponding to the outside air while a corpse in a basement will largely assume a temperature related to the soil temperature. To estimate the microclimate of a corpse found in either area, we need answers to the following questions:

1. Is the basement conditioned or unconditioned?
2. Is the basement finished or unfinished?
3. Was the door to the basement closed or open?
4. What fraction of the wall height is below-grade?
5. What is the floor and wall composition?

Crawlspaces

Some buildings have gaps between the floor and the ground (crawlspaces). Crawlspaces can be classified as sealed with plastic over the soil floor, closed with no exchange of air with the outside, and open with air exchange with the outside either using ventilation (mechanically-ventilated) or infiltration (naturally-ventilated). Crawlspaces are found under both constructed-in-place and manufactured homes and are influenced by the outside air, the ground, and the overlying building (Figure 6.7).

Exchange rates of the crawlspace under natural infiltration with the outside air average 4.6 per hour.[91] The average air temperature of naturally-ventilated crawlspaces in cold climates were similar to the outside air during the summer, about 9 °F warmer in the fall and about 18 °F warmer in winter and spring.[92] Mechanically-ventilated crawlspaces a cold climate have similar temperatures as the outside air during the summer, 2 °F warmer during fall, and 4 °F and 11 °F warmer during the winter and spring.[92] Crawlspace air temperatures in a coastal climate average 10 °F warmer than the outside air.[91] A naturally-ventilated residential crawlspace in a temperate climate during the winter and spring showed air temperatures vary more than an adjoining unfinished conditioned basement (Figure 6.7a). The average air temperature in this crawlspace, typically 19 °F above that of the outside air during the winter, changed 0.3 °F per 1 °F change in the outside air.[64]

A naturally-ventilated crawlspace in a temperate climate ranged from 50-60% in the winter to 95-98% in the summer.[93] The RH of a mechanically-ventilated air in the crawlspace of a cold climate decreased with increasing circulation from 83% to as low as 65%.[92]

The ground is often the primary source of humidity. An unventilated crawlspace RH is typically 85% to 95%.[94] Covering the soil floor reduced the RH of this crawlspace to 74% while the air over the ground varied from 85% to 100%.[92] However, in maritime climates the infiltrating air may contribute more humidity than the ground. Crawlspaces in a maritime climate had an average RH of 65% with air temperatures above the **dew point temperature** less than 1% of the time.[95] Closing off outside infiltration in this crawlspace decreased the RH to 58% – indicating the outside maritime air was more humid than the soil.[95] A crawlspace in a drier environment had an average

RH of 64% with no periods of 100% RH.[95] Closing off the infiltration in this crawlspace did not change the humidity in the crawlspace.[95]

Crawlspaces of manufactured homes are typically skirted. If a manufactured house is not skirted, the humidity under the home will be the same as the outside air. A survey of manufactured houses found air temperatures in skirted crawlspaces averaged of 8.8 °F warmer than the outside air.[96] A survey of houses in a hot moist climate showed that a solid skirt resulted in dew point temperatures 5 °F to 7 °F above the outside air while a perforated skirt resulted in dew point temperatures 2 °F above the outside air.[97]

A corpse lying on the ground in a crawlspace will be influenced by ventilation-affected air temperatures in the space and the temperature of the ground. To estimate the microclimate of the crawlspace, we need answers to the following questions:

1. Is the crawlspace conditioned or unconditioned?
2. What are the crawlspace walls composed of?
3. Is the crawlspace ventilated? How?
4. What is the gap area for air infiltration with the outside air

Buildings with passive heating and cooling

Unconditioned buildings include barns, detached garages, cabins, and storage buildings. The indoor air temperature involves an energy budget including solar radiation receipt on the roofs and walls, thermal radiation from the walls and roof, conduction of heat through the roof and walls, infiltration of outside air, and conduction of heat between the floor and the air in building are the dominant factors in the air temperature (Figure 6.7).

Solar radiation heats the outer walls and varies by exposure and composition. A sunlit south-exposure exterior wall of brick in the northern hemisphere was 11 °F warmer than the outside air while a similarly exposed red wood wall was 15 °F warmer.[98] The corresponding sunlit east-exposure walls were 19 °F warmer and 24 °F warmer for the brick and red wood siding.[98] In an unheated building in the northern hemisphere with 3-ft-thick stone walls, a north-facing room always had air temperatures the same as the outdoors while a south-facing room had air temperatures 9 °F higher than the outside air in the fall and winter and about 5 °F higher in the summer.[99]

Indoor air temperatures are typically less variable than the outside air due in part to the **insulation** and heat storage of the walls and ceilings. Insulation will however reduce temperature variations. A simulation showed that when the outside air temperatures varied by 95 °F, the indoor air temperature of an unheated unventilated building with brick walls and concrete ceilings varied by 50 °F while one using insulating mineral wool on the wall and ceiling varied by only 10 °F.[100]

The storage of heat within a building and its contents also reduces temperature variations. An unheated residence with wood construction had average air temperatures indoors that were 2 °F to 5 °F warmer in the fall and early winter than the outside and 10 °F to 12 °F warmer during the late winter and spring.[101] The inside air temperatures of a 5-ft thick limestone wall building varied from 27 °F to 68 °F while outside air temperatures varied from −2 °F to 68 °F.[102]

Air infiltrates through gaps in windows, walls, doors and the roof. An unheated manufactured home had infiltration rates varying between 0.1 and 0.6 per hour, increasing by 0.2 per hour for every 18 °F in indoor-outdoor temperature difference.[103] Infiltration of a 5-ft thick limestone wall building was 0.3 per hour if doors and windows were closed and 0.6 per hour if open.[102] Garages can have widely varying rated of infiltration with the outside air depending on the opening and windspeed. Garages with main doors on the windward side had exchange rates of 1.1 per hour if closed and 19 per hour if open under a 3 to 5 mph wind and 13 per hour if closed and 66 per hour if open when under 13 mph wind.[104]

Because hot air rises, warm air will rise up stairwells and air at the same height on different floors of a house will likely differ. A lodge in open land in maritime climate increased air temperature by 3.3 °F per story[101] while that for a temperate climate residence in a woodland increased 1.7 °F per floor.[64]

Porous wall materials such as unpainted plaster, stone, and brick damp outside moisture variations.[105] The humidity in an unheated building with 3-ft-thick stone walls had a north-facing room with an average RH the same as the outside air and a south-facing room with about 25% higher RH than the outside.[99] The RH in an unheated residence in a maritime climate averaged 60% to 70% during the fall, winter and spring while another averaged 40% to 60% during the fall and winter and 30% to 35% in spring.[101] The RH averaged 90% to 87% depending on location in an unheated building with 5-ft thick limestone walls.[102]

Unconditioned buildings usually have either concrete floors or bare ground. Consequently, the ground can be a major source of humidity and sink for heat. However, if infiltration is high, the humidity may be close to the outside air. Over time the soil surface temperature under the building will tend towards the average temperature in the overlying building.

Since most corpses in buildings are found on the floor, heat exchange with the floor is an important energy exchange pathway, especially since the floors are usually bare ground or concrete. Air flow in the building can be moderate requiring consideration of convective heat exchange between the ground, corpse and air. The corpse will approach the room air temperature over time. See Appendix Cases 3, 4 and 6 for estimates of air temperatures in unheated buildings.

To estimate the microclimate of the corpse in an unconditioned building, we need baseline environment information including:

1. Composition of roof, walls, floors
2. Window area on each wall
3. Window/door area of open windows/doors
4. Volume and location of room where corpse is found
5. Volume of the house by floor
6. Whether door of room where corpse is found is closed or open

Buildings with active heating and cooling

Conditioned buildings include most residences and office buildings as well as cabins using space heaters. The conditioning of interior spaces in buildings (controlling the minimum or maximum air temperatures) introduces the additional challenge of assessing the contribution of the heating and/or air conditioning (AC) on the air.

The climate in any given room will depend on its size, position in the building, exposure (compass direction), condition of any windows, infiltration of air into and out of the room through doors, and composition of floor, wall and ceiling. Heating can be through a central system that distributes heat throughout the building or by space heaters in specific rooms. The ability of added heat to warm the air in a room will depend on the relative contribution of the heater's heat input to the other sources and sinks of heat: window solar radiation penetration, window and door infiltration, and floor, wall and ceiling heat conduction. Since heating the air also heats everything else in the room, you need to have an idea of the heat storage of the room's floor, walls, and ceiling as well as contents.

Ventilation or infiltration of outside air varies widely. A survey of homes (both manufactured and in-place construction) showed 59% of the air circulating in a house comes from the outside of the house.[106] Manufactured homes have infiltration rates of about 0.1 per hour,[96] double-wide manufactured home had infiltration rates between 0.2 and 0.5 per hour,[103] and in-place construction homes have average infiltration rates of 0.5 per hour with higher rates in urban areas (0.9 per hour).[107] Infiltration rates differ depending on if the heat or AC is on: the nominal air exchange with the outside of a 1500 square foot house is 0.8 per hour when heating and 0.4 per hour when cooling.[107]

Fan-driven mechanical ventilation systems use the difference between the temperature of the air in the room with the thermostat controller and thermostat temperature setting to control ventilation. The percentage of time that the AC was on in a double-wide manufactured home increased by 40% for every 10 °F difference in inside versus outside air temperature.[103] The percentage of time that the heater was on in that same house increased by 20% for every 18 °F difference in inside versus outside air temperature.[103] During the summer, the time that air conditioning was on in a temperate climate woodland residence increased 32% for each 10 °F increase in outside temperature above the thermostat temperature.[64] During the winter, that same residence increased the time the heater was on by 100% for each 10 °F decrease in outside temperature under thermostat temperature.[64]

As discussed above, air temperatures on different floors of a house can be expected to differ. For a 2-story residence in a temperate climate woods, the air temperature increased an average of 1.7 °F per story when heated during the heating season and 3.7 °F per story when cooled during the cooling season.[64]

The RH of conditioned spaces is nominally 30% to 60%.[107] An increase in **water vapor** with height was found for a maritime climate lodge with an average RH of 33% on the 1st floor and 42% on the third floor while the air temperatures increased with floor.[101]

Natural convection of heat from a corpse, driven by buoyancy, is usually much greater than **forced convection** since air flow in a room is generally very low. Since most corpses are found on the floor, heat exchange with the floor is an important energy exchange pathway. A radiative balance will develop between the corpse and the walls and ceiling such that over time the corpse temperature will closely approximate the air temperature in the room. See Appendix Case 5 and Case 7 for estimates of air temperatures in heated buildings.

To estimate the microclimate of the corpse in a conditioned building, we need a large amount of baseline information including:

1. Temperature setting of thermostat
2. Does thermostat control heating/cooling
3. Whether thermostatic control of heating/cooling is on
4. Type of heating of room
5. Window area on each wall
6. Window/door area of open windows/doors
7. Composition of outer wall and floor
8. Floor or story of the building the corpse is found
9. Volume and location of room where corpse is found
10. Volume of the house by floor
11. Whether door of room where corpse is found is closed or open

References

1. Mutiibwa, D., Strachan, S. and Albright, T., 2015. *IEEE Journal of Selected Topics in Applied Earth Observations and Remote Sensing*, *8*(10), pp.4762–4774.
2. Xu, M., Chen, J. and Brookshire, B.L., 1997. *Climate Research*, *8*(3), pp.209–223.
3. Oke, T.R., 1978. *Boundary Layer Climates*. Methuen and Co., NY, NY, 372p.

4. Joy, J.E., Liette, N.L. and Harrah, H.L., 2006. *Forensic Science International, 164*(2–3), pp.183–192.
5. Huntington, T.E., Higley, L.G. and Baxendale, F.P., 2007. *Journal of Forensic Sciences, 52*(2), pp.453–458.
6. Charabidze, D., Bourel, B. and Gosset, D., 2011. *Forensic Science International, 211*(1–3), pp.61–66.
7. Heaton, V., Moffatt, C. and Simmons, T., 2014. *Journal of Forensic Sciences, 59*(3), pp.676–682.
8. Johnson, A.P. and Wallman, J.F., 2014. *Forensic Science International, 241*, pp.141–149.
9. Mann, RW, Bass, W.M. and Meadows, L. 1990. *Journal of Forensic Sciences 35*, 1: 103–11.
10. Goff, M. L., 2009. *Experimental and Applied Acarology, 49*, pp.21–36.
11. Lee, M.J., Voss, S.C., Franklin, D. and Dadour, I.R., 2018. *Forensic Science International, 289*, pp.175–185.
12. Slone, D.H. and Gruner, S.V., 2007. *Journal of Medical Entomology, 44*(3), pp.516–523.
13. Deonier, C.C., 1940. *Journal of Economic Entomology, 33*(1).
14. Chapman, B., Raymer, C. and Keatley, D.A., 2022. *Homicide Studies, 26*(2), pp.199–215.
15. Geiger, R., 1965. *Climate near the ground*. Harvard University Press, Cambridge, MA, USA, 611p.
16. Rorison, I.H., Sutton, F. and Hunt, R., 1986. *Plant, Cell & Environment, 9*(1), pp.49–56.
17. Häkkänen, H., Hurme, K. and Liukkonen, M., 2007. *Journal of Investigative Psychology and Offender Profiling, 4*(3), pp.181–197.
18. Sea, J. and Beauregard, E., 2018. *International Journal of Offender Therapy and Comparative Criminology, 62*(7), pp.1947–1966.
19. Ritchie, J.T. and Nesmith, D.S., 1991. *Modeling Plant and Soil Systems, 31*, pp.5–29.
20. Maddonni, G.A. and Otegui, M.E., 1996. *Field Crops Research, 48*(1), pp.81–87.
21. Moon, K., Duff, T.J. and Tolhurst, K.G., 2019. *Fire Safety Journal, 105*, pp.320–329.
22. Abd Latif, Z. and Blackburn, G.A., 2010. *International Journal of Biometeorology, 54*, pp.119–129.
23. Baldocchi, D. and Ma, S., 2013. *Tellus B: Chemical and Physical Meteorology, 65*(1), p.19994.
24. Haesen, S., Lembrechts, J.J., De Frenne, P., Lenoir, J., Aalto, J., Ashcroft, M.B., Kopecký, M., Luoto, M., Maclean, I., Nijs, I. and Niittynen, P., 2021. *Global Change Biology, 27*(23), pp.6307–6319.
25. Renaud, V., Innes, J.L., Dobbertin, M. and Rebetez, M., 2011. *Theoretical and Applied Climatology, 105*, pp.119–127.
26. Finklin, A.I. 1986. *A climatic handbook for Glacier National Park- with* data *for Waterton Lakes National Park*. Forest Service, United States Dept. of Agriculture, Intermountain Experiment Station Gen. Tech. Rpt. INT-204. 124p.
27. Brooks, R.T. and Kyker-Snowman, T.D., 2008. *Forest Ecology and Management, 254*(1), pp.65–73.
28. Souch, C.A. and Souch, C., 1993. *Journal of Arboriculture, 19*(5), pp.303–312.

29. Sharratt, B.S., Baker, D.G., Wall, D.B., Skaggs, R.H. and Ruschy, D.L., 1992. *Agricultural and Forest Meteorology*, *57*(4), pp.243–251.
30. Fu, Q., Hou, R., Li, T., Wang, M. and Yan, J., 2018. *Geoderma*, *325*, pp.9–17.
31. Magni, P.A., Petersen, C., Georgy, J. and Dadour, I.R., 2019. *Global Journal of Forensic Sci & Med* 1 (3): 2019. *GJFSM. MS. ID*, *513*.
32. Janaway, R.C., Percival, S.L. and Wilson, A.S., 2009. *Microbiology and aging: clinical manifestations*, pp.313–334.
33. Landsberg, H.E. and Blanc, M.L., 1958. *Soil Science Society of America Proceedings*, *22*, pp.491–495.
34. Penrod, E.B. and Stewart, O.W., 1967. *Transactions of the American Society of Agricultural Engineers*, *10*(3), pp.359–0363.
35. Smith, G.D., Newhall, F., Robinson, L.H. and Swanson, D. 1964. US Department of Agriculture, Soil Conservation Service *SCS-TP-166*, 14p.
36. Rodriguez, W.C. and Bass, W.M., 1985 *Journal of Forensic Sciences*, *30*(3), pp.836–852.
37. Schneider, M. and Goss, K.U., 2012. *Geoderma*, *170*, pp.64–69.
38. Archer, J.R. and Smith, P.D., 1972. *Journal of Soil Science*, 23(4), pp.475–480.
39. Micozzi, M.S., 1986. *Journal of Forensic Sciences*, *31*(3), pp.953–961.
40. Webb, B.W. and Zhang, Y., 1997. *Hydrological Processes*, *11*(1), pp.79–101.
41. Rodgers, G.K. and Anderson, D.V., 1961. *Journal of the Fisheries Board of Canada*, *18*(4), pp.617–636.
42. Gallego-Elvira, B., Baille, A., Martín-Górriz, B. and Martínez-Álvarez, V., 2010. *Hydrological Processes: An International Journal*, *24*(6), pp.758–766.
43. Rutherford, J.C., Marsh, N.A., Davies, P.M. and Bunn, S.E., 2004. *Marine and Freshwater Research*, *55*(8), pp.737–748.
44. Ham, J., Toran, L. and Cruz, J., 2006. Effect of upstream ponds on stream temperature. *Environmental Geology*, *50*, pp.55–61.
45. Moore, R.D., Spittlehouse, D.L. and Story, A., 2005. *Journal of the American Water Resources Association*, *41*(4), pp.813–834.
46. Webb B, Hannah D, Moore R, Brown L, Nobilis F. 2008. *Hydrological Processes 22*, pp.902–918.
47. Caissie, D., 2006. *Freshwater Biology*, *51*(8), pp.1389–1406.
48. Benz, S.A., Bayer, P. and Blum, P., 2017. *Environmental Research Letters*, *12*(3), p.034005.
49. Stefan, H.G. and Preud'homme, E.B., 1993. *Journal of the American Water Resources Association*, *29*(1), pp.27–45.
50. Bogan, T., Mohseni, O. and Stefan, H.G., 2003. *Water Resources Research*, *39*(9).
51. Woolway, R.I., Jones, I.D., Maberly, S.C., French, J.R., Livingstone, D.M., Monteith, D.T., Simpson, G.L., Thackeray, S.J., Andersen, M.R., Battarbee, R.W. and DeGasperi, C.L., 2016. *PLOS one*, *11*(3), p.e0152466.
52. Kettle, A.J., Hughes, C., Unazi, G.A., Birch, L., Mohie-El-Din, H. and Jones, M.R., 2012. *Journal of Hydrology*, *470*, pp.12–27.
53. McCombie, A.M., 1959. *Limnology and Oceanography*, *4*(3), pp.252–258.
54. Naumenko, M.A. and Guzivaty, V.V., 2022. *Geography and Natural Resources*, *43*(1), pp.69–76.
55. Wallace, J.R. and Merritt, R.W., 2020. pp. 155–186 In: Byrd, J.H. and Tomberlin, J.K. (Eds.) *Forensic Entomology: The utility of arthropods in legal investigations.* 3rd Ed, CRC Press, Taylor and Francis Group, Baca Raton, FL, USA.

56. Boutelier, C., Bougues, L. and Timbal, J., 1977. *Journal of Applied Physiology*, *42*(1), pp.93–100.
57. Abraham, J., Cheng, L., Vallez, L. and Wei, T., 2022. *Journal of Forensic Sciences*, *67*(3), pp.1049–1059.
58. Garratt, J.R., 1990. *Boundary-Layer Meteorology*, *50*, pp.171–203.
59. Gregory, S and Smith, K. 1967. *Weather* 22, pp. 497–505.
60. Raynor, G.S., 1971. *Forest Science*, *17*(3), pp.351–363.
61. Rosenberg, N.J., Blad, B.L., Verma, S.R. 1983. *Microclimate: the biological environment*. 2nd Ed., John Wiley and Sons, NY, NY 495p.
62. Chen, J., Saunders, S.C., Crow, T.R., Naiman, R.J., Brosofske, K.D., Mroz, G.D., Brookshire, B.L. and Franklin, J.F., 1999. *BioScience*, *49*(4), pp.288–297.
63. Davies-Colley, R.J., Payne, G.W. and Van Elswijk, M., 2000. *New Zealand Journal of Ecology*, pp.111–121.
64. Grant, R.H., 2024. Unpublished measurements
65. Rambo, T.R. and North, M.P., 2008. *Northwest Science*, *82*(4), pp.259–268.
66. Anderson, P.D., Larson, D.J. and Chan, S.S., 2007. *Forest Science*, *53*(2), pp.254–269.
67. Segal, M., Leuthold, M., Arritt, R.W., Anderson, C. and Shen, J., 1997. *Bulletin of the American Meteorological Society*, 78, 1135–1147.
68. Scott, R.W. and Huff, F.A., 1996. *Journal of Great Lakes Research*, *22*(4), pp.845–863.
69. Atkinson, B.W. 1981. *Meso-scale Atmospheric circulations*. Academic Press, London. 495p.
70. Al-Qahtni, A.H., Mashaly, A.M., Alajmi, R.A., Alshehri, A.A., Al-Musawi, Z.M. and Al-Khalifa, M.S., 2019. *Saudi Journal of Biological Sciences*, *26*(7), pp.1499–1502.
71. Anderson, G.S. 2020. pp. 103–139 In: Byrd, J.H. and Tomberlin, J.K. (Eds.) *Forensic Entomology: The utility of arthropods in legal investigations*. 3rd Ed, CRC Press, Taylor and Francis Group, Baca Raton, FL, USA.
72. Dadour, I.R., Almanjahie, I., Fowkes, N.D., Keady, G. and Vijayan, K., 2011. *Forensic Science International*, *207*(1–3), pp.205–211.
73. Scott, K.I., Simpson, J.R. and McPherson, E.G., 1999. *Journal of Arboriculture*, *25*(3), pp.129–142.
74. Driscoll, D.M. personal communications. *Preliminary report, The car as solar collector*. Texas A&M University, College Station, TX, USA.
75. Rashid, R.M., 2018. *Thermal management of vehicle interior temperature for improvement of fuel economy*, Doctoral dissertation, University of Windsor, Canada.
76. Byard, R.W., 2023. *The American Journal of Forensic Medicine and Pathology*, pp.10–1097
77. Singh, S.P., Saha, K., Singh, J. and Sandhu, A.P.S., 2012. *Packaging Technology and Science*, *25*(3), pp.149–160.
78. Moreau, G., Lutz, L. and Amendt, J., 2021. *Pure and Applied Geophysics*, *178*, pp.745–756.
79. Magni, P.A., Dhaliwal, S.S., Dadour, I.R. 2016. *Journal of Medical Entomology*, *53*(4), pp.782–789.
80. Bhadra, P., Hart, A.J. and Hall, M.J.R., 2014. *Forensic Science International*, *239*, pp.62–72.
81. Heisler, G.M. 1986. *Urban Ecology*, *9*, pp.337–359

82. Larsen, L., Gronlund, C.J., Ketenci, K.C., Harlan, S.L., Hondula, D.M., Stone Jr, B., Lanza, K., Mallen, E., Wright, M.K. and O'Neill, M.S., 2023. *Journal of the American Planning Association, 89*(3), pp.363–375.
83. Mazarron, F.R. and Canas, I., 2008. *Energy and Buildings, 40*(10), pp.1931–1940.
84. Kononova, M.S., Zherlykina, M.N. and Kononov, A.D., 2021, *IOP Conference Series: Materials Science and Engineering* (Vol. 1079, No. 3, p. 032061). IOP Publishing.
85. Emery, A.F., Heerwagen, D.R., Kippenhan, C.J. and Steele, D.E., 2007. *Science and Technology for the Built Environment, 13*(1), p.39.
86. Gryc, A., Was, K. and Radon, J., 2011. *Infrastruktura i Ekologia Terenów Wiejskich*, (12).
87. Kacálek, P. and Petříček, T., 2018. In *MATEC Web of Conferences* (Vol. 146, p. 03003). EDP Sciences.
88. Nawalany, G. and Sokołowski, P., 2019. *Energies, 12*(15), p.2922.
89. Armstrong, M.M., Ruest, K. and Swinton, M.C., 2011. *Assessing the imp-sact of cold climate on basement temperatures.* Canadian Centre for Housing Technology.
90. Hoffmann, C., Geissler, A., Huber, H. and Hauri, C., 2019. In *Journal of Physics: Conference Series* (Vol. 1343, No. 1, p. 012180). IOP Publishing.
91. Francisco, P.W. and Palmiter, L., 2007. *American Society of Heating, Refrigeration, and Air Conditioning Engineers Transactions, 113*(2), pp.81–90.
92. Kurnitski, J., 2000. *Energy and Buildings, 32*(1), pp.19–39.
93. Biswas, K., Christian, J. and Gehl, A., 2011. Oak Ridge National Laboratory, NOAA, Department of Commerce. *ORNL/TM-2011/438, October.*
94. Kurnitski, J., 2001. *Building and Environment, 36*(3), pp.359–373.
95. Hales, D., Lubliner, M. and Gordon, A., 2010. Vented and Conditioned Crawlspace Performance in Marine and Cold Climates of the Pacific Northwest. *ASHRAE Buildings XI Conference.*
96. Davis, B., Siegel, J. and Palmiter, L., 1996. *Proceedings of the 1996 Summer Study on Energy Efficiency in Buildings, "Profiting from Energy Efficiency"*, American Council for an Energy Efficient Economy (ACEEE), USA, Washington DC, AIVC #10268, 61–71.
97. Beal, D. and Chasar, D., 2006. *Proceedings, 15th Symposium on Improving Building systems in hot and humid climates.* ESL-HH-06-07-42.
98. Landsberg, H. 1954. *Meteorological Monographs*, 2(8), pp.81–98.
99. Ryhl-Svendsen, M., Padfield, T., Smith, V.A. and De Santis, F., 2003. pp. 7–11 In *Healthy Buildings 2003, Proceedings of ISIAQ 7th International Conference.*
100. Bøhm, B. and Ryhl-Svendsen, M., 2011. *Energy and Buildings, 43*(12), pp.3337–3342.
101. Tibbetts, D.C. and Robson, D.R., 1963. *Temperature and relative humidity in park buildings maritime provinces.* National Research Council of Canada, Division of Building Research, Ottawa, Canada, Internal Rpt. 268, 12p.
102. Napp, M. and Kalamees, T., 2015. *Energy and Buildings, 108*, pp.61–71.
103. Nabinger, S. and Persily, A.K., 2008. *Airtightness, ventilation, and energy consumption in a manufactured house: Pre-retrofit results.* US Department of Commerce, National Institute of Standards and Technology, NISTIR #7478, 27p.
104. Alevantis, L.E. and Girman, J.R., 1989. *Proceedings of IAQ'89, The Human Equation, Health and Comfort*, San Diego, CA (1989), pp. 184–191

105. Padfield, T and Larsen, P.K., 2004. *Studies in Conservation*, *49*(2), pp.131–137.
106. Francisco, P.W. and Palmiter, L., 1996. *Proceedings of the 1996 Summer Study on Energy Efficiency in Buildings, "Profiting from Energy Efficiency"*, American Council for an Energy Efficient Economy (ACEEE), Washington DC, USA, AIVC #10270, 103–115
107. ASHRAE. 2001. *2001 Fundamentals, SI edition*. ASHRAE, Inc, Atlanta GA, USA.

Sources of and problems with climate records

7

The entomological approach to determining **PMI** uses **temperature** information at the body location while the **taphonomical** approaches to PMI determination require temperature and sometimes **humidity** information. Measurements of air temperature and humidity are rarely available where a body is found. The problem then is how to estimate the air temperature and **RH** where the body was found from places where the air temperature and humidity are measured and **record**ed. And of course, for the information to be useful at the location of the body, we have to have confidence in the recorded information. So, what are the best measurement record sources to make the best estimate or guess at what the conditions were where the body was found?

Where can I find useful measurement records?

There are many **climate** and **weather** networks across the USA as well as across the world that make routine measurements of air temperature and humidity. These networks can be grouped into two major categories: those that report the daily maximum and minimum temperatures and **precipitation** for a day and those that report temperatures, winds, humidity and other variable either hourly or more frequently through a day.

Sources and errors of US daily measurement records

Cooperative (**COOP**) 'Summary of the Day' stations measure daily maximum, minimum, and current air temperatures and precipitation accumulations. There is no measurement of humidity made at these stations. These stations are located in **open terrain**, qualified and maintained by the National Weather Service (**NWS**), and operated by private individuals. There are about 8500 stations across the US. This network is the primary source of daily climate records in the USA. Measurements are made at a height of 4 ft to 6 ft above the ground (agl). Until 1988, observations at these COOP stations were manually recorded on the 'Record of River and Climatological Observations' form (WS Form B-91) from daily observations of liquid-in-glass thermometers (Figure 7.1) in

DOI: 10.4324/9781003486633-7

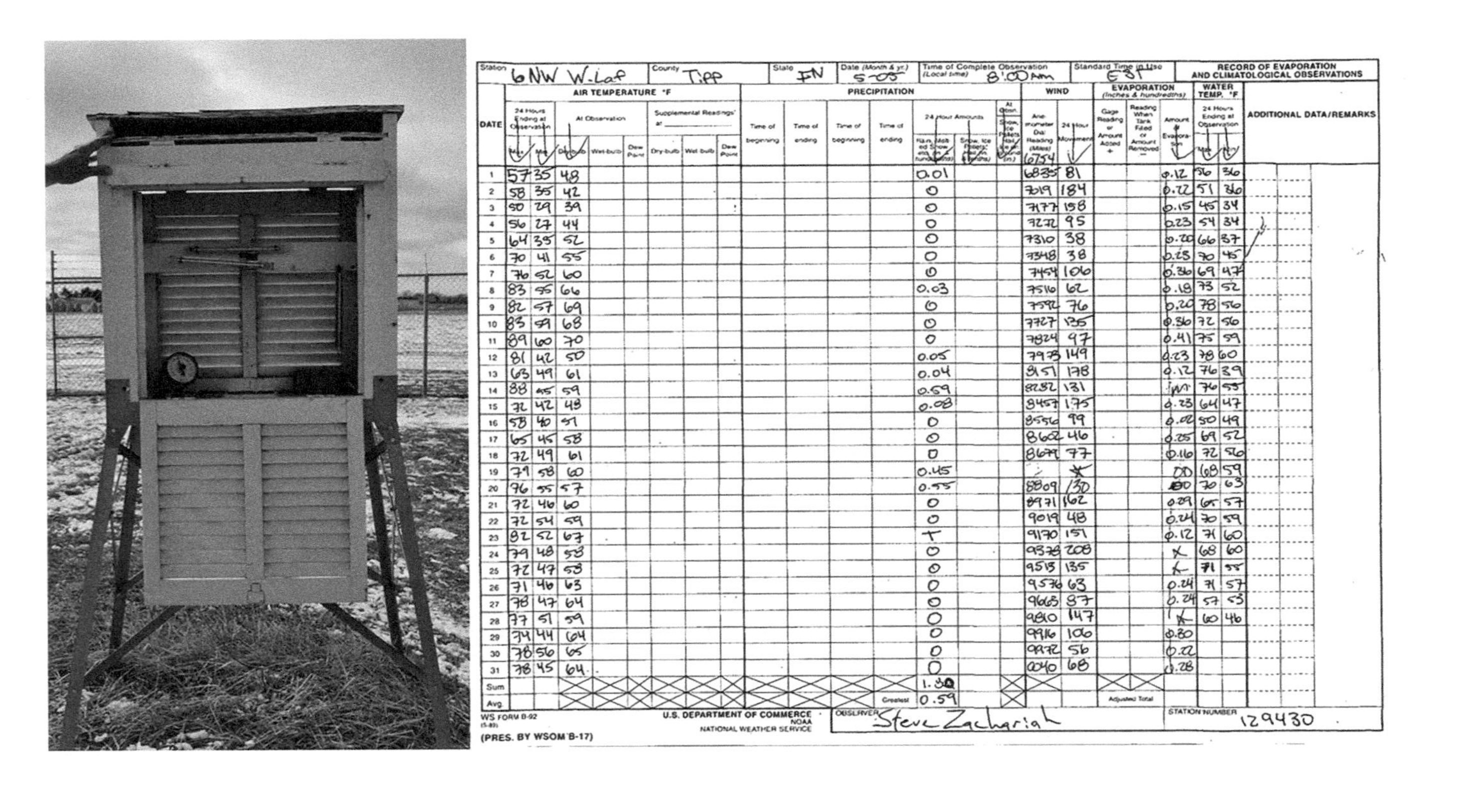

Station: 6 NW W-Laf | County: Tipp | State: IN | Date (Month & yr.): 5-05 | Time of Complete Observation (Local time): 8:00 Am | Standard Time in Use: EST

RECORD OF EVAPORATION AND CLIMATOLOGICAL OBSERVATIONS

DATE	AIR TEMPERATURE °F: 24 Hours Ending at Observation, Max	Min	At Observation: Dry-bulb	PRECIPITATION: 24 Hour Amounts, Rain, Melted Snow, etc. (in & hundredths)	WIND: Anemometer Dial Reading (Miles)	24 Hour Movement	EVAPORATION (Inches & hundredths): Amount of Evaporation	WATER TEMP. °F: 24 Hours Ending at Observation, Max	Min
					6754				
1	57	35	48	0.01	6835	81	0.12	56	36
2	58	35	42	0	7019	184	0.22	51	36
3	50	29	39	0	7177	158	0.15	45	34
4	56	27	44	0	7272	95	0.23	54	34
5	64	35	52	0	7310	38	0.20	66	37
6	70	41	55	0	7348	38	0.25	70	45
7	76	52	60	0	7454	106	0.36	69	47
8	83	55	66	0.03	7516	62	0.19	73	52
9	82	57	69	0	7592	76	0.20	78	56
10	83	59	68	0	7727	135	0.36	72	56
11	89	60	70	0	7824	97	0.41	75	59
12	81	42	50	0.05	7973	149	0.23	78	60
13	63	49	61	0.04	8151	178	0.12	76	39
14	88	45	59	0.59	8282	131	[illegible]	76	55
15	72	42	48	0.08	8457	175	0.23	64	47
16	58	40	51	0	8556	99	0.02	50	49
17	65	45	58	0	8602	46	0.25	69	52
18	72	49	61	0	8679	77	0.16	72	56
19	79	58	60	0.45		*	[illegible]	68	59
20	76	55	57	0.55	8809	130	[illegible]	70	63
21	72	46	60	0	8971	162	0.29	65	57
22	72	54	59	0	9019	48	0.24	70	59
23	82	52	67	T	9170	151	0.12	71	60
24	79	48	53	0	9378	208	X	68	60
25	72	47	53	0	9513	135	X	71	55
26	71	46	63	0	9576	63	0.24	71	57
27	78	47	64	0	9663	87	0.24	57	53
28	77	51	59	0	9810	147	X	60	46
29	74	44	64	0	9916	106	0.30		
30	78	56	65	0	9972	56	0.22		
31	78	45	64	0	0040	68	0.28		
Sum				1.30					
Avg.			Greatest	0.59					

WS FORM B-92 (5-89) (PRES. BY WSOM B-17)

U.S. DEPARTMENT OF COMMERCE NOAA NATIONAL WEATHER SERVICE

OBSERVER: Steve Zachariah

STATION NUMBER: 129430

Figure 7.1 COOP 'Summary of the Day' Cottom Region temperature shelter with liquid-in-glass thermometers and typical manually-recorded Form B-91 record.

Photograph by author R.H. Grant. Form image source: NOAA National Center for Environmental Information https://www.ncei.noaa.gov/access

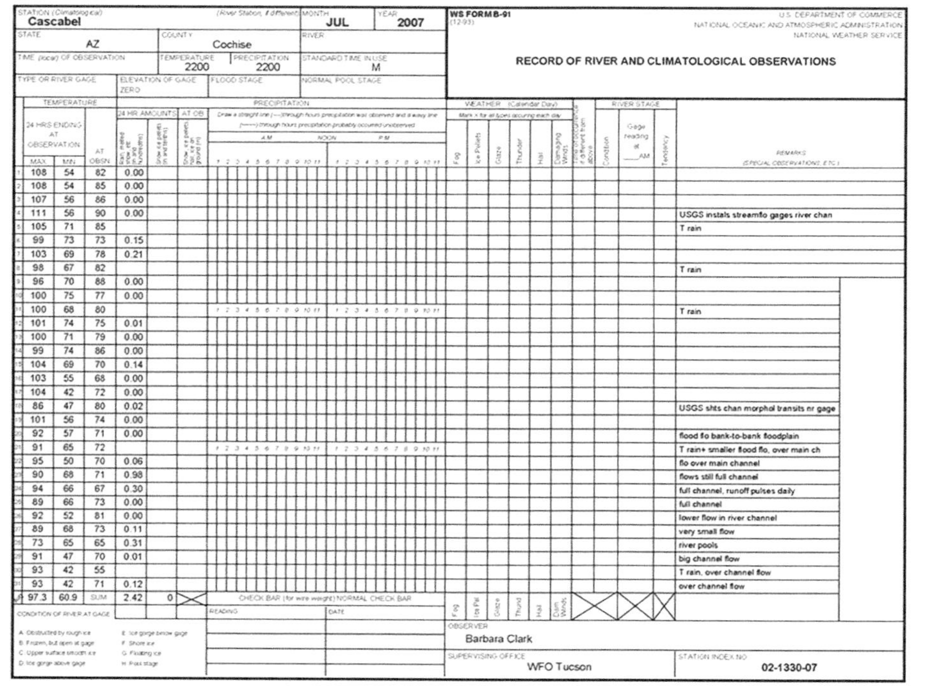

STATION (Climatological): Cascabel | MONTH: JUL | YEAR: 2007

STATE: AZ | COUNTY: Cochise | RIVER:

TIME (local) OF OBSERVATION: TEMPERATURE 2200 | PRECIPITATION 2200 | STANDARD TIME IN USE: M

TYPE OF RIVER GAGE | ELEVATION OF GAGE ZERO | FLOOD STAGE | NORMAL POOL STAGE

WS FORM B-91 (12-93)

U.S. DEPARTMENT OF COMMERCE
NATIONAL OCEANIC AND ATMOSPHERIC ADMINISTRATION
NATIONAL WEATHER SERVICE

RECORD OF RIVER AND CLIMATOLOGICAL OBSERVATIONS

Day	Temperature MAX	MIN	AT OBSN	Precipitation 24 HR AMOUNTS	Snow	Remarks (Special observations, etc.)
1	108	54	82	0.00		
2	108	54	85	0.00		
3	107	56	86	0.00		
4	111	56	90	0.00		USGS instals streamflo gages river chan
5	105	71	85			T rain
6	99	73	73	0.15		
7	103	69	78	0.21		
8	98	67	82			T rain
9	96	70	88	0.00		
10	100	75	77	0.00		
11	100	68	80			T rain
12	101	74	75	0.01		
13	100	71	79	0.00		
14	99	74	86	0.00		
15	104	69	70	0.14		
16	103	55	68	0.00		
17	104	42	72	0.00		
18	86	47	80	0.02		USGS shts chan morphol transits nr gage
19	101	56	74	0.00		
20	92	57	71	0.00		flood flo bank-to-bank floodplain
21	91	65	72			T rain+ smaller flood flo, over main ch
22	95	50	70	0.06		flo over main channel
23	90	68	71	0.98		flows still full channel
24	94	66	67	0.30		full channel, runoff pulses daily
25	89	66	73	0.00		full channel
26	92	52	81	0.00		lower flow in river channel
27	89	68	73	0.11		very small flow
28	73	65	65	0.31		river pools
29	91	47	70	0.01		big channel flow
30	93	42	55			T rain, over channel flow
31	93	42	71	0.12		over channel flow
	97.3	60.9	SUM	2.42	0	

CONDITION OF RIVER AT GAGE

OBSERVER: Barbara Clark

SUPERVISING OFFICE: WFO Tucson

STATION INDEX NO: 02-1330-07

Figure 7.2 COOP 'Summary of the Day' MMTS radiation shield with thermistor and typical automated Form B-91 record.

Photograph by author R.H. Grant. Form image source: NOAA National Center for Environmental Information https://www.ncei.noaa.gov/access

a wooden **radiation** shelter (to minimize **solar radiation** and **thermal radiation** influences on the measurements).[1]

After 1988 the temperature measurement systems were automated with measurements made using an electronic Maximum/Minimum Temperature Systems (**MMTS**) is housed in a much smaller radiation shield to minimize radiation influences on the measured temperature (Figure 7.2). The daily maximum and minimum temperature observations as well as the current temperature are read by the operator and manually recorded in a web-based **database**.[2] The precipitation accumulation is manually measured daily from a 7-inch diameter rain gauge and prior to 1998 was manually recorded on the WS Form B-91 (Figure 7.1) but is now added to the web-based database and automatically incorporated into the WS Form B-91 (Figure 7.2). While the manually-prepared WS Form B-91 show when precipitation occurred, the computer-generated forms do not show this. You can determine the type of **sensor** used at a given station in a given year by checking the pdf of the actual B-91 forms on the National Centers for Environmental Information website (www.climate.gov/maps-data/dataset/cooperative-observers-network-observation-forms-data-tables) or inspecting the **metadata** for the site at www.ncei.noaa.gov/access/homr/.

The air temperature measured by liquid-in-glass maximum thermometers in the wooden shelter overestimate the true temperature by 1.5 °**F** while the liquid-in-glass minimum thermometer in the shelter underestimated the true temperature by 1.0 °F.[3] The daily average temperature, based on the liquid-in-glass maximum and minimum thermometers in the wooden shelter, is +/− 1.8 °F (BOX 7.1) assuming it is on a typical clear sky day. The combined **error** of the MMTS sensor and radiation shield (Figure 7.2) is +/− 1.1 °F for the minimum value and 1.0 °F for the maximum value.[3] The daily average temperature error, calculated from the maximum and minimum temperature, is +/− 1.5 °F assuming a typical clear sky day.

SENSOR OR SENSOR SYSTEM ACCURACY

Each sensor/sensor system is produced to fall within a certain bracket to the true temperature (as defined by a **NIST**-certified sensor). The sensor/sensor system accuracy, stated as '+/−' some value, indicates the variability in the sensor/sensor system measurement. For all values reported in this chapter, the accuracy is describing the range of values for which 95% of all sensors/sensor systems can be expected to record given the actual value.

Sources and errors of US hourly measurement records

The approximately 900 station climatological/meteorological measurement network composed of the Automated Surface Observation Systems (**ASOS**) and Automated Weather Observation Systems (**AWOS**) are designed to primarily support aviation operations and weather forecasting (Figure 7.3) and are calibrated and maintained by the NWS. ASOS stations make measurements every minute, 24 hours a day, every day of the year. ASOS stations measure ambient pressure, air temperature, **dew point temperature**, wind speed and direction, **cloud height** and amount (clear, scattered, broken, overcast) up to 12,000 feet and precipitation accumulation. AWOS stations are located at local and regional airports and may not measure at all hours and every day of the year. Furthermore, AWOS stations may not make as many measurements as the ASOS stations: all AWOS stations measure ambient pressure, air temperature, dew point temperature, wind speed and direction but only AWOS III stations provide cloud height and amount and precipitation accumulation.

- The hourly air temperature and dew point temperature sensor for all ASOS/AWOS stations is measured within a ventilated radiation shield mounted at 4.5 ft **agl** (Figure 7.3). Hourly temperature records are 5-minute averages just before the hour of 3- or 5-second measurements. Wind speed and direction is measured at 27 ft or 33 ft agl.[2] The ASOS/AWOS dew point temperature sensor has an **accuracy** of between 1.1°F and 4.7°F with greater error at temperatures below freezing (between 3.1°F and 7.9°F).[2] The temperature/humidity

Figure 7.3 Typical ASOS weather station. Temperature sensor is mounted in a radiation shield at the far end of the instrumentation rail.

Photograph by author R.H. Grant.

sensor installed in the standard radiation shield has an error of +/– 1.8 °F [4] and an average RH error estimated to be +/– 4% (assuming a range of air and dew point temperatures from freezing to 86 °F).

- The ASOS/AWOS wind measurements are made at 27 ft agl. Recorded hourly wind speed and direction are averages taken within 15 minutes of the hour by observation and recorded manually prior to 1996 and are now automatically calculated 2-minute averages of 5 second measurements.[2] The ASOS/AWOS wind measurements are automatically recorded with a wind speed accuracy of +/– 2.3 mph and wind direction accuracy of +/– 5 deg.[2]
- Cloud base heights and the sky cloud that the covers are measured directly over the airport using a laser beam ceilometer every 30 seconds and averaged over 30 minutes (NWS, 2018). Cloud height estimates are automatically recorded and have an accuracy of +/– 50 ft.[2] **Cloud cover** is reported in five categories (0 to 5%, 5% to 25%, 25% to 50%, 50% to 87%, and 87% to 100%) for clouds lower than 12,000 ft agl.[2] Metadata for each station, including what is measured and how, is available for all ASOS or AWOS stations at www.ncei.noaa.gov/access/homr/.

Personal weather stations (PWS), available through many weather station manufacturers and distributors, are growing in popularity worldwide (Figure 7.4). In the USA, the most common PWS network is established with the Weather Underground Company.[5] Their PWS network, with over 250,000 stations worldwide, has no network coordination- with stations included in the Weather Underground network on personal choice. There are no qualifications for station siting or maintenance to be part of the network- many stations are located in backyards and are never recalibrated after purchase. Measurements typically include air temperature and RH at around 5 ft **agl** and wind speed and direction at over 7 ft agl. Radiation shields for the Temperature/RH sensors are encouraged but not required. Measurements are automatically recorded and can be uploaded into the web-based database. There is no metadata for the Weather Underground PWS. Generally, there is no site-specific metadata available for any PWS. There are often estimates of measurement errors for these networks based on information from a subset of stations.

Mesoscale stations (**MesoNets**) are typically in relatively dense networks within a region and are generally an excellent source of information. Among other variables, these stations usually measure air temperature, RH, barometric pressure, soil temperature, solar radiation, wind direction and speed, and accumulated precipitation (Figure 7.4). Winds are typically measured at 6 to 10 ft agl and air temperature and humidity are generally measured at 4.5 ft. agl. Air temperature and humidity are usually measured using a capacitive

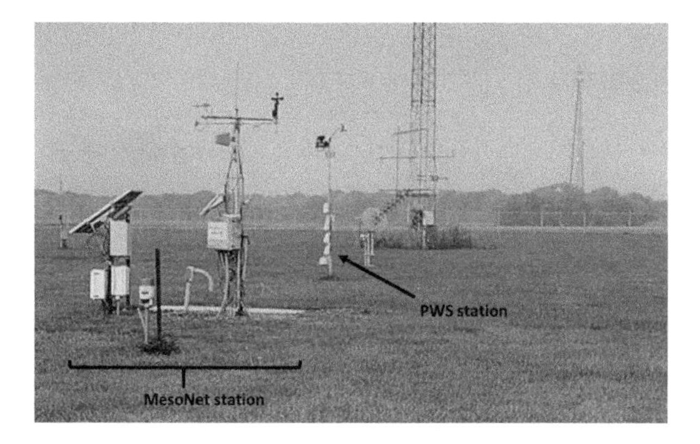

Figure 7.4 Typical MesoNet station and PWS in an open grass area.

Photograph by author R.H. Grant.

hygrometer coupled with a thermistor or platinum resistance thermometer. The errors in the air temperature and RH measurement are typically 0.7 °F and +/− 3% respectively, but the error depends on the specific sensor and radiation shield used in each network.[6] The error in wind speed and direction likewise are sensor specific but typically +/− 1.1 mph and +/− 5° respectively. The best way to find a MesoNet station, information on sensors used at a station, and climate records for the station is to contact either the US State Climatologist in the state where the body was found or contact the appropriate US Regional Climate Center that serves that state (www.ncei.noaa.gov/regional/regional-climate-centers).

Climate Reference Network (**CRN**) stations are found spread widely across the US is large open terrain (144 stations in the US). They each measure air temperature and RH at 4.5 ft agl., solar radiation, surface temperature, accumulated precipitation, and surface wind speed at 33 ft agl. height. In addition, each station measures soil temperature and moisture at 5 depths.[7] The air temperature sensor contained within the radiation shield has a measurement error of +/−0.7 °F.[4] There are three temperature sensors on each station, so averaging the three measurements together will result in an air temperature error of +/−0.4 °F. The error in RH is +/− 3% while that for wind speed and direction is +/− 0.6 mph.[8] These records are available from the **NCEI** website www.ncei.noaa.gov/products/land-based-station/us-climate-reference-network. Metadata including what is measured and how is available at www.ncei.noaa.gov/access/homr/.

Stations in Roadway Weather Information Systems (**RWIS**) can be found on major US highways and are generally maintained by an individual state's Departments of Transportation (DOT). They typically measure air temperature, dewpoint temperature, wind speed and direction, accumulated precipitation, barometric pressure, as well as a wide variety of road conditions. Although the sensors used and mounting heights of the sensors are likely consistent across a given state's network, there are no standards for siting, maintenance and calibration. The US Department of Transportation guidelines state that RWIS wind measurements should be made at 33 ft agl and air temperature and humidity measurements should be made at 5 to 6.5 ft agl.[9] Metadata for US RWIS stations may be available through the specific state DOT.

Other US sources of climate information

There are a wide variety of sources for other environmental measurements that may be useful in estimating a given **local climate**. These may include (but not be limited to) solar radiation measurements, lake or ocean water and air temperature measurements, river levels, weather maps, snow cover, and river levels.

Solar radiation measurements or estimates may be needed to model the **microclimate** at the corpse location. Solar radiation measurements are commonly made at MesoNet stations. Several small solar radiation measurement networks exist that monitor conditions with sub-hourly **resolution**, however they are so widely spaced that they are unlikely to be useful for any specific location of interest (the SOLRAD network has nine stations across the US while BSRN has six). Estimates of daily solar radiation in a given area are also available from satellite measurements[10] (www.ngdc.noaa.gov/stp/solar/solarirrad.html).

Snow cover information may be needed to determine if the corpse is under snowpack at the time of discovery or possibly under snowpack during part of the PMI. Most COOP stations report snow depth. Snow cover information is provided from satellites. The satellite-based snow cover maps (www.nohrsc.noaa.gov/nsa/) provide higher spatial resolution than the COOP stations can provide – especially in mountainous areas.

Water temperatures may be needed if a corpse is partially or entirely immersed in water bodies. Lake water and air temperatures are often made by municipalities that draw water from freshwater lakes such as used in Appendix Case 7. Temperatures and wind conditions near some shorelines are measured by oceanic or lake buoys. Recent and historical buoy measurements can be downloaded from the **NOAA** National Data Buoy Center website (www.ndbc.noaa.gov/). River water temperatures are available for many rivers and streams from the US Geological Survey water quality monitoring stations[11] at waterwatch.usgs.gov/wqwatch/. River levels may be needed to assess possibility of bodies being only partly in the water or only in water part of the postmortem time. These **data** can be found at waterdata.usgs.gov/nwis/rt.

Soil temperatures may be needed if the bodies are buried. Regionally, soil temperatures at depths to 40" are available for the north central US from the North Central River Forecast Center at www.weather.gov/ncrfc/LMI_SoilTemperatureDepthMaps and for the western US from the United States

Department of Agriculture Natural Resources Conservation Service at www.nrcs.usda.gov/resources/data-and-reports/soil-climate-analysis-network. Soil temperatures to depths of 40" at CRN stations are available from the National Centers for Environmental Information (NCEI) at www.ncei.noaa.gov/access/crn/qcdatasets.html.

Non-US sources of daily or hourly measurement records

Daily minimum, average, and maximum air temperatures and cumulative precipitation at climate stations and air temperature, dew point temperature, RH, barometric pressure, winds, cloud cover and height at hourly airport stations in Canada are available from the Meteorological Service of Canada at climate.weather.gc.ca/historical_data/search_historic_data_e.html. This, and other national airport weather networks have similar measurement heights and accuracies to those of the US ASOS/AWOS network.

There are daily minimum, average, and maximum air temperature and precipitation records, (as well as humidity, barometric pressure, solar radiation, wind speed, wind direction, cloud cover, and snow depth) for weather and climate stations in 65 countries across Europe and the Mediterranean. These are available through the European Climate Assessment and Dataset (**ECAD**) website www.ecad.eu/dailydata/index.php. However, many countries do not allow weather and climate records to be downloaded (knmi-ecad-assets-prd.s3.amazonaws.com/documents/ECAD_datapolicy.pdf). Most countries recognize the importance of the World Meteorological Organization (**WMO**) sensor guidelines[12, 13] and strive to equip their networks accordingly. These WMO guidelines indicate that air temperature sensors should have an error of at most +/− 0.4 °F, dew point temperatures should an error of at most +/− 0.5 °F, RH an error of at most +/− 3%, cloud base height an error of at most 33 ft, wind speed an error of at most 1 mph, and wind direction an error of at most +/− 5°. RWIS stations are found on roads in many countries across the world. Similarly, there are many PWS networks in non-US countries with similar measurements mentioned above for US PWS networks. However, the station siting, sensor mounting, sensors used, and sensor calibration and maintenance (with corresponding sensor error estimates) need to be well understood before the records should be used.

Are you confident in the validity of the measurement record?

Confidence in the quality of the measurements has both quality control and quality assurance aspects. Aspects of measurement quality control of particular import include the siting or location of the station, the accuracy of the sensor system making the measurement, and the representativeness of the station location to the location of interest. Most of this information is found in the 'metadata' for the station. Our quality assurance evaluation includes that the sensor is performing within specifications and the station location still meets siting criteria.

Station siting issues

You generally want to use a station that is and has been in an open grassy terrain and then adjust the record using various modeling approaches (Chapter 9) to account for the differences of conditions between where the measurements were made and where the body was found. Generally, stations are sited in an area of unobstructed view of the sky and unobstructed wind (Figure 7.4). Most measurement station networks have established specific siting criteria designed to make the measurements most useful for the purpose of the network. The World Meteorological Organization (WMO) siting guidelines are widely used across the world.[12, 14] These guidelines indicate that: 1) the stations should represent areas of around 60 mi by 60 mi and 2) the area around the proposed site should not differ within a 1600 ft radius, and 3) airport weather stations should be sited to represent conditions on the runways.

The siting criteria for establishing and maintaining COOP stations include being at least 100 ft from any paved of concrete surface, be in an open clearing no closer than four times the height of the nearest obstruction (tree, building, fence, etc). Siting on building roofs is allowed with specific approval. Unfortunately, these criteria are not consistently followed.[15]

Each MesoNet network across the US has their own siting criteria for the initial installation and the adherence to the criteria are documented in the metadata. They typically include: 1) minimizing influences of urban structures and roads, irrigation, forests and large water bodies, 2) minimizing slope of the site and surrounding landscape, 3) minimizing surrounding obstructions that limit wind flow and **sky view**.[14]

The advised siting criteria for the US Weather Underground® PWS are similar to those established for COOP stations, but the fulfillment of the criteria is neither documented or verified. Wind sensors are advised to be the highest object around, at 33 ft above agl if possible and preferably on a roof top.

AWOS and ASOS stations are located in the grassy centers of airports and more than 100 ft from the runway. The ASOS/AWOS measurements are automatically recorded at airports using sensors located either in the center of the airport between runways or at the edge of the runway. Consequently, a given airport station may not be useful to the PMI determination.

Likewise, CRN stations are not likely to change over time since one of the criteria for locating them is the likelihood for long-term stability of the surrounding landscape.[8] These stations are located to represent a regional climate and individual measurements at the stations are assessed as to spatial representativeness[16] and generally at least 30 ft from any vegetation or buildings.

Guidelines in the USA for RWIS station siting vary by type of station: regional or local.[9] Regional stations are ideally located 20 to 30 miles apart along the highway and on relatively flat terrain upwind of the predominant wind direction. The guidelines indicate the sensor tower should be 30 to 50 ft from the roadway. The surrounding vegetation should be 'low vegetation' within 50 ft of the tower. Local stations are located to represent conditions resulting from changes in topography or roadway type.

Metadata associated with the European and Mediterranean climate stations is also available through the European Climate Assessment and Dataset (ECAD) website www.ecad.eu/dailydata/index.php.

Even when the initial siting criteria of a station is met, over time the environment around the station may change. Changes that affect the local climate of the station have been shown to dramatically change the US COOP station measurement record.[1] Even moves of ASOS/AWOS stations within an airport can result in **bias**es in the air temperature and dew point temperatures relative to other locations.[17] The history of a station's sensors and location are documented in the station metadata. Preferably the actual station should be visited and photographed to verify the quality of the station siting. GoogleEarth® can also be used to assess any building development near the station of interest.

Time of daily recorded values

There are time-of-observation issues in determining average daily temperatures based on daily minimum and maximum temperature measurements. The time when a daily minimum and maximum temperature measurements are recorded at COOP stations influences how to best use that record: maximum and minimum temperature measurements are typically made in the early morning. However, a typical clear sky day has the minimum air temperature occurring just before sunrise and the maximum temperatures occurring in the prior day afternoon (Figure 7.5). Consequently, the recorded maximum temperature is typically shifted to represent the prior day. If there is a single day of missing observations, this shift will result in additional days of incomplete daily observations. If the measurements are recorded in the afternoon the maximum and minimum temperature are likely to represent the same day as the record indicates. This 'time of observation correction' over can be substantial[18] and must be considered prior to using the COOP station records. The observation time is indicated in the metadata for a given station as well as on the Form B-91s (Figures 7.1 and 7.2).

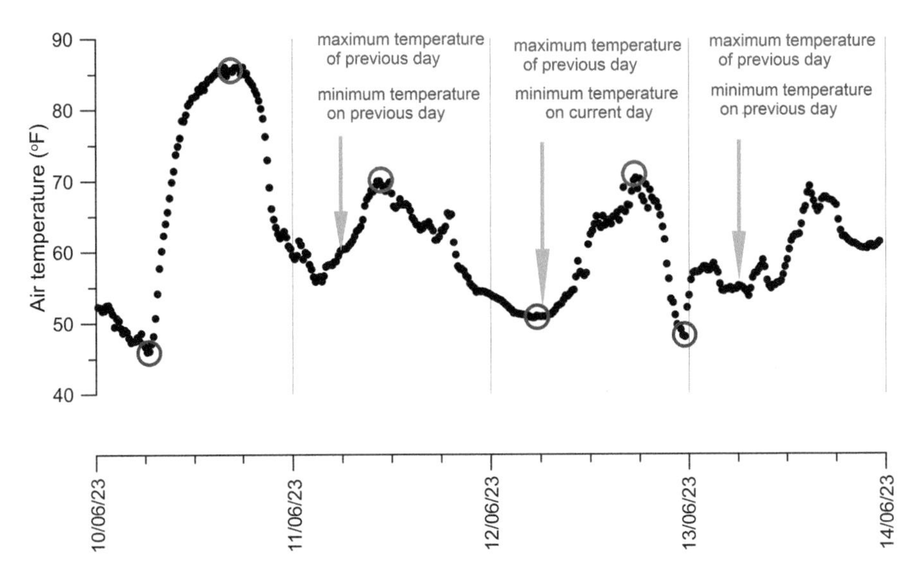

Figure 7.5 Time of Observation problem. Multiple days illustrating the time of occurrence of daily maximum and minimum temperature compared to what is reported at 6AM at a COOP station (text) for the observation time (green arrows).

Assessment of measurement errors

How good is the measurement record? Checks should include various internal measurement value checks, comparison of values to known sensor failure mode values, continuity of the time series of measurements, similarity of values to those reported in prior years at the same time of year, and similarity of measurement values with surrounding stations.

A comparison of the temperature at the time of observation to the recorded maximum and minimum temperatures provides a basic check of the validity of the observations at COOP stations. A daily maximum and minimum temperature record is invalidated if the recorded temperature at the time of observation did not fall within the range of recorded maximum and minimum daily temperatures. If both air and dew point temperatures are available (ASOS/AWOS and many RWIS stations), an internal check on the validity of the record is to verify that the air temperature was less than or equal to the dew point temperature. ASOS/AWOS dew point sensor failures are indicated by repeated 32 °F values.[2] If both air temperatures and RH are reported (MesoNet and CRN stations), the RH values need to be converted to dew point temperatures before an evaluation of the humidity measurement is possible. Once converted, an internal check is made to verify that the dew point temperature is nearly constant over the course of most days. The chance of having a humidity measurement fail the quality assurance checks is much greater than air temperatures. RH sensors are more likely to fail the air temperature or dew point temperature sensors. Measurements from the AWOS, ASOS, CRN, and some RWIS networks are part of the Meteorological Assimilation Data Ingest System[19] (madis-data.ncep.noaa.gov/) and have been quality assured using a wide range of quality assurance criteria. PWS networks have no calibration or maintenance protocol established to assure measurement record quality. As a result, PWS records are generally not used in the estimating air temperature and/or humidity conditions at a location of body discovery. If a PWS is critically important for a case, WMO guidelines should be used to assess sensor errors and measurement record accuracy.[20]

Daily and hourly measurement records from Canadian stations are assimilated into their National Climate Archives.[21] These records are not gap-filled but quality assurance tests of reasonableness have been made. The 'non-blended' data from the ECAD are as received from the participating country and have already had quality assurance checks conducted by their respective network protocols. The 'blended data' includes quality assurance tests for changes in observation time, and reasonableness of the daily temperature range (maximum-minimum temperatures), day-to-day change in daily temperature range and daily average RH.

How can you estimate values for missing records?

A complete temporally sequential set of hourly or daily measurements is needed to determine averages of air temperature and RH or accumulations of heat over time. Often the preferred climate station record is missing a few measurements. What is the best way to estimate these missing measurements?

- Missing daily air temperature records can be estimated using various interpolation or substitution methods depending on the availability and suitability of measurement records at the station of interest and nearby stations.
- Missing humidity or dew point temperature records are uncommon but the approach to estimating missing values is identical to that for air temperatures after the RH values are converted to **specific humidity**.

Temporal interpolation at the same station

Estimating missing records at one station when there are no other stations in the area requires an understanding of the weather conditions (Figure 7.6) on the days with the missing records. Daily weather maps showing the location of weather **front**s can be downloaded at the NWS Weather Prediction Center website (www.wpc.ncep.noaa.gov/dwm/dwm.shtml).

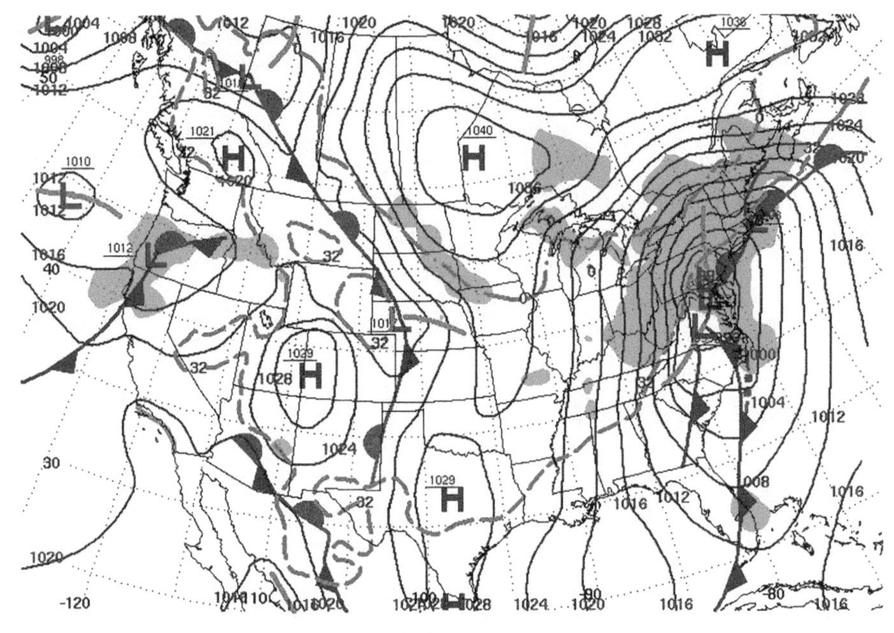

Figure 7.6 A Daily Weather Map illustrating the patterns of weather for a given day at 7AM EST. High and Low Pressure centers are indicated with a blue 'H' or red 'L' respectively. Regions with precipitation are indicated by green. Areas of relatively low pressure, called 'troughs', are indicated by dashed orange lines while cold and warm fronts are indicated by blue lines with triangles and red lines with half circles respectively.

Source: NOAA NWS Weather Prediction Center, National Center for Environmental Prediction: www.wpc.ncep.noaa.gov/dwm/dwm.shtml

If there is no frontal passage at the station with missing records, then a direct temporal (or time) interpolation across the missing days may be possible (Figure 7.7). It is important to recognize that the accuracy of this method decreases as the number of sequential days of missing values increases- especially when inspection of the frontal weather systems shows fronts moving through the area where the station is located.

Mean difference between station records

Estimating missing records using records from nearby stations with similar environments first requires an evaluation of large-scale weather systems during the missing days- for example are the weather systems on opposite sides of weather fronts. Comparisons of station records over time will often 'capture' the evidence of a warm or cold front passing through the area of interest.

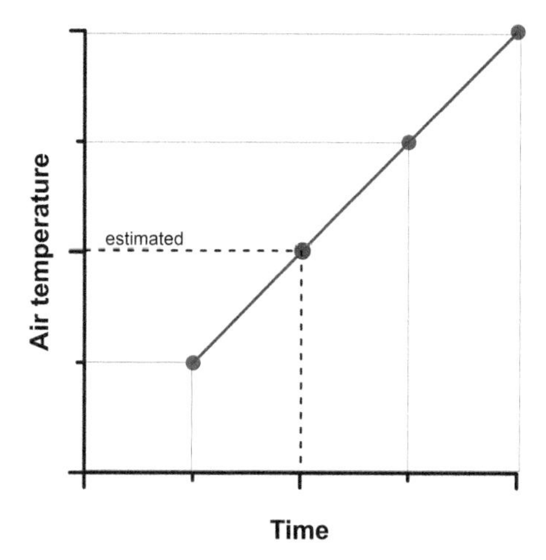

Figure 7.7 Estimating a missing value in the temporal record. The temperature for the missing time is estimated by linear interpolation (blue line) of recorded values (blues circle) on days before and after (red circle) the missing time.

Extreme differences in minimum of maximum daily temperatures between two or more stations on some days but not most days may be due to cold or warm fronts present between the two stations.

If there are no fronts between potentially useful nearby stations with valid records and the station of interest with missing records, differences in station sensors and siting must be considered. Such differences often cause differences in a recorded measurement under identical conditions. 'Summary of the Day' COOP network MMTS temperature sensors generally have warmer (+0.9 °F) daily maximum and cooler (−0.7 °F) minimum temperatures than the CRN sensors with considerable variability.[22] 'Summary of the Day' COOP stations averaged 0.6 °F warmer maximum temperatures and 0.5 °F cooler minimum temperatures than typically equipped MesoNet stations.[22] Another comparison of COOP and MesoNet stations showed maximum temperature differences to be +/− 10 °F and differences in both the daily average and minimum temperatures of +/− 5 °F.[24]

Acceptable differences in measurements from stations in different networks can be determined by evaluating the differences in the record when both stations are measuring air temperature and differences in the network air temperature sensor system accuracy. Here we will consider only air temperature measurements but a similar analysis would need to be made for humidity measurements. Let's assume the difference in the daily minimum air temperature at two stations was +/− 1 °F over the period that both stations had valid records. If sensor1 at station1 had and accuracy of +/− 1 °F and sensor2 at station2 had an accuracy of +/− 1.8 °F we would conclude that the two station's records represented measurements of the same conditions. A difference of up to 2.8 °F between the station records might be possible while the two sensors are measuring the identical conditions. Such consistency in representation of the, in this case, temperature would give confidence that records from either location could be used to fill-in missing data of the other location. However directly substituting the missing observation with that of another location introduces variability that is only a result of the substitution.

If there are only two stations with useable measurement records and the microclimates of the two stations are essentially identical, we can determine the relationship between the measurement records of the two stations when there are valid measurements made at both locations (called paired observations) and use the relationship to estimate the missing record. If the interpolation is for maximum and minimum air temperatures and the two stations vary in altitude by more than about 100 ft, the reported temperatures should be adjusted for differences in altitude (see Chapter 4) before a spatial interpolation.

Assume we have missing values in the *station1* record and need to estimate the missing records with adjusted values from *station2*. While we might expect the air temperatures to vary over time based on the passages of weather systems and changes in season, the differences between two stations will likely be similar. Even if the air temperatures do not have a **normal distribution**,[25] the difference between the air temperatures of two station is likely to be nearly normal.[26]

The mean difference or 'bias' in estimating the value at one station from another is called the mean bias error or **MBE**. This mean 'offset' in the air temperature estimate varies with each pair of values. A statistical description of the variation in paired differences between the values of one station and another station at the same time is called the 'root mean squared error' or **RMSE**. The RMSE statistic assumes that the paired differences in air temperature between the two stations are normally distributed.[27] The average difference between the station values (MBE) is then used to adjust the value to be substituted in to the longer record with the record gap. The accuracy of the estimate is represented by the RMSE. High MBE indicates the two stations differ substantially in their microclimates. High RMSE indicates the microclimate differ in their response to changes in the local climate and changes in the large-scale weather. This methodology was used in Appendix Case 8.

Linear regression between station records

If the environments around the two stations differ substantially or there are weather front passing through during the period of record, we do not expect the value at *station 2* to be equal to that at *station 1* for any given day even when the same sensors mounted in the same way are being used. We can calculate how the difference in environment, location and sensors at the two stations influences the values. Assuming the air temperature at *station1* and

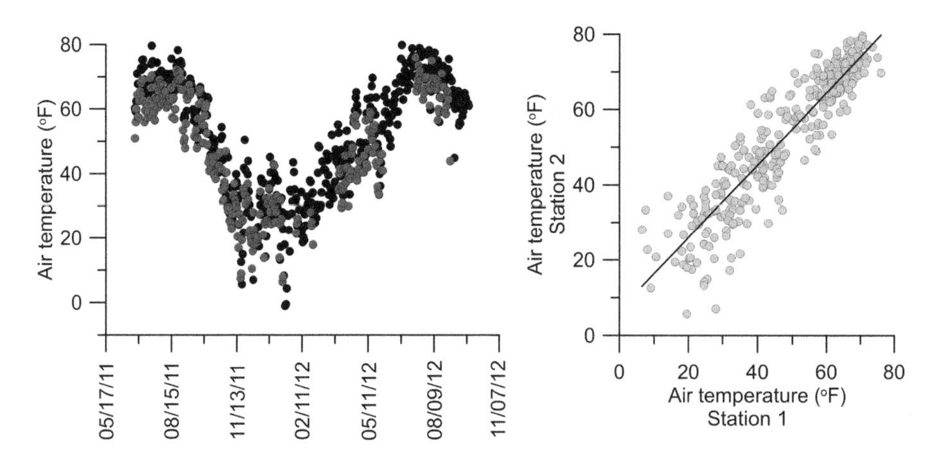

Figure 7.8 Estimating the missing daily values at *Station 2* from *Station 1* values over a period of time. Called 'Linear Regression Least Squares Method'.

Data source: NOAA National Center for Environmental Information.

station2 are not responding to the very same environmental conditions, we can 'regress' the paired values (valid values from *station 2* and *station 1* over a period of time) using a least-squares method[28] (Figure 7.8). Given the least squares equation, we can estimate the missing values for station 1 given the measurement record for station 2 on those days. The typical accuracy measure of this estimation, called the '**coefficient of determination**', assumes normally-distributed air temperatures- which is generally not a valid assumption for air temperature or humidity values. As in Figure 7.8, differences between station air temperatures are often greater at temperatures below freezing due to high spatial variability in snow depth and cover.

Spatial interpolation from other station records

If there are many stations near the station where you need to estimate missing values, the distance between this station and the other stations can be used to 'weight' the usefulness or applicability of measurements at a nearby station in estimating a missing value(s) at a station or location using a method called 'inverse distance weighting'.[29] For example, in Figure 7.9 a station closer to the desired location '?' and in a similar microclimate would be considered more accurate and 'weighted' more than one further away and in a similar microclimate. Station 'D' is most distant from station '?' and differs in environment and probably local climate influences (see Chapter 4) while stations 'A', 'B', and 'C' are similar in environment and altitude from station '?'. Of these stations, station B is closest to station'?' and the measurement records at the station would be "weighted" more than stations A and C in estimating

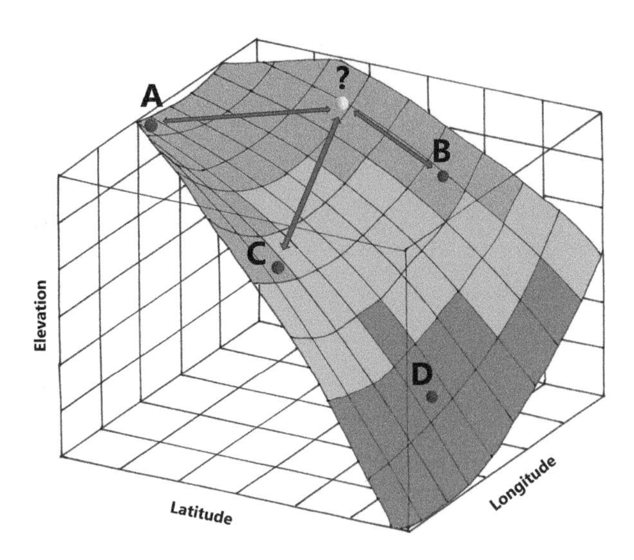

Figure 7.9 Spatial Interpolation. Estimating missing values based on averaging multiple station values base on distance from the station (red circles) with the missing values (blue circle). Stations labelled 'A' through 'D'. Arrows represent distance between a station and the station with an unknown value (?). Rocky ground, grassland, and forest are indicated by brown, light green and dark green.

the values at Station '?'. If the interpolation is for daily maximum and minimum air temperatures and the stations vary in altitude by more than about 100 ft, the reported temperatures should be adjusted for differences in altitude (see Chapter 4) before spatial interpolation. Stations chosen for this method should have similar environments and not be located on opposite sides of a weather front during the missing days of record. An advantage of this approach is the reduction in the average daily temperature error estimate since it is based on multiple values (see Chapter 10). An example of this methodology is in Appendix Case 1.

Limitations

If the missing air temperature or humidity records are hourly values, it is generally better to estimate the values using temporal rather than spatial interpolation. This is because the hourly weather conditions may very over short periods of time across typical station spacing due to lags associate with terrain type, vegetation, and slope orientation.

The European and Mediterranean climate station records available through the ECAD website fill daily record gaps with observations from nearby stations, provided that they are within 12.5 km distance and that height differences are less than 25 m.

How can you get a legally-admissible copy of the climate record?

Submission of weather or climate data as evidence in US or Canadian court may require certification that the data record has not been altered.[30] Certification in the USA can be made for Daily Weather Maps representing the weather at 7AM EST, Local climatological Data (3-hourly weather observations of temperature, RH, precipitation, cloudiness, wind direction/speed at ASOS/AWOS stations), Climatological Data (daily maximum and minimum temperatures and precipitation at COOP stations with some stations also including daily snowfall, snow depth, **evaporation** and soil temperature data), COOP station data (daily maximum and minimum temperatures and precipitation) and COOP station monthly B-91 forms, ASOS surface weather data, and Buoy data. This certified data can be submitted directly to the court or subpoena the record from NCEI. It is important to remember that the certified data is the original data as reported (COOP station data, Buoy data and ASOS/AWOS data) and does not include any quality assurance checks or substitution of data, does not assure observer expertise or specific sensor accuracy, and does not attest to the conditions under which the measurements were recorded.

Certification of hourly or daily weather and climate records in Canada can be requested by email to the Climate Services Office within the region of interest (climate.weather.gc.ca/contactus/climate_services_e.html).

References

1. Changnon, S.A. and Kunkel, K.E., 2006. *Journal of Atmospheric and Oceanic Technology*, *23*(6), pp.825–828.
2. NWS. 2018. **NWS** *Instruction 10-1302*. 28p.
3. Hubbard, K.G.; X. Lin and E. Walter-Shea. 2001. *Journal of Atmospheric and Oceanic Technology* 18:851–864.
4. Lin, X., Hubbard, K.G., Walter-Shea, E.A., Brandle, J.R. and Meyer, G.E., 2001. *Journal of Atmospheric and Oceanic Technology*, *18*(9), pp.1470–1484.
5. Weather Underground. 2023. *Personal Weather Network. Installing your personal weather station*. www.wunderground.com/pws/installation-guide (Accessed 2 October, 2023
6. Erell, E., Leal, V. and Maldonado, E., 2005. *Boundary-Layer Meteorology*, *114*, pp.205–231.
7. Bell, J.E., Palecki, M.A., Baker, C.B., Collins, W.G., Lawrimore, J.H., Leeper, R.D., Hall, M.E., Kochendorfer, J., Meyers, T.P., Wilson, T. and Diamond, H.J., 2013. *Journal of Hydrometeorology*, *14*(3), pp.977–988.
8. Diamond, H.J., Karl, T.R., Palecki, M.A., Baker, C.B., Bell, J.E., Leeper, R.D., Easterling, D.R., Lawrimore, J.H., Meyers, T.P., Helfert, M.R. and Goodge, G., 2013. *Bulletin of the American Meteorological Society*, *94*(4), pp.485–498.

9. FHA. 2005. Federal Highway administration, Department of Transportation. *Pub. FHWA-HOP-05-026.*
10. NSRDB. 1992. *User's Manual, National Solar radiation data base (1961-1990),* Version 1 National Renewable Energy Laboratory, Department of Energy. 144p.
11. Wagner, R.J., Boulger Jr., R.W., Oblinger, C.J., Smith, B.A., 2006. USGS, U.S. Department of the Interior, *A Techniques and Methods* 1-D3. 91p.
12. Murthy, V. R. K.; Grant, R.; Milford, J.; Oliphant, A.; Orlandini, S.; Stigter, K. and Wiering, J. 2010. pp. 1–35 In *Guide to Agricultural Meteorological Practices-WMO-No. 134,* World Meteorological Organization.
13. WMO. 2021a. *Guide to Instruments and Methods of Observation, Vol. I-Measurement of Meteorological Variables,* WMO 8, WMO, Geneva, Switzerland.
14. WMO. 2021b. *Guide to Instruments and Methods of Observation, Vol. III-Observing systems,* WMO 8, WMO, Geneva, Switzerland.
15. Davey, C.A. and Pielke Sr, R.A., 2005. *Bulletin of the American Meteorological Society, 86*(4), pp.497–504.
16. NOAA. 2002. *Climate Reference Network (CRN) Site information handbook.* National Oceanic and Atmospheric Administration, National Environmental Satellite, Data, and Information Service, NOAA/NESDIS CRN Series X030
17. Guttman, N.B. and Baker, C.B., 1996. *Bulletin of the American Meteorological Society, 77*(12), pp.2865–2874.
18. Janis, M.J., 2002. *Journal of Applied Meteorology and Climatology, 41*(5), pp.588–603.
19. MADIS. 2023. *Meteorological Assimilation Data Ingest System.* National Centers for Environmental Prediction, National Oceanic and Atmospheric Administration. madis.ncep.noaa.gov/index.shtml (accessed 1 October, 2023).
20. Peltier, R.E.; Castell, N.; Clements, A.L.; Tim Dye, T.; Hüglin, C.; Kroll, J.H.; Lung, S.-C. C.; Ning, Z.; Parsons, M.; Penza, M.; Reisen, F. and von Schneidemesser, E. 2021. *WMO 1215,* WMO, Geneva, Switzerland.
21. ECCC. 2021. *MANOBS Manual of surface weather observation standards.* 8th Ed., Amendment 1, December 2021. Environment and Climate Change Canada.
22. Leeper, Ronald D., Jared Rennie, and Michael A. Palecki. 2015. *Journal of Atmospheric and Oceanic Technology* 32(4), pp.703–721.
23. Holder, C., Boyles, R., Syed, A., Niyogi, D. and Raman, S., 2006. *Journal of Atmospheric and Oceanic Technology, 23*(5), pp.671–682.
24. Wu, H., Hubbard, K.G. and You, J., 2005. *Journal of Atmospheric and Oceanic Technology, 22*(5), pp.592–602.
25. Harmel, R.D., Richardson, C.W., Hanson, C.L. and Johnson, G.L., 2002. *Journal of Applied Meteorology and Climatology, 41*(7), pp.744–753.
26. Kale, B.K., 1970. *The American Mathematical Monthly, 77*(9), pp.992–995.
27. Hodson, T.O., 2022. *Geoscientific Model Development, 15*(14), pp.5481–5487.
28. Brooks, C.E.P. and Carruthers, N., 1953. *Handbook of statistical methods in meteorology.* Air Ministry, Meteorological Office, M.O. 538. Her Majesty's Stationary Office, London. 412p.
29. Panigrahi, N. 2023. Inverse Distance Weight. In: Daya Sagar, B.S., Cheng, Q., McKinley, J., Agterberg, F. (eds) *Encyclopedia of Mathematical Geosciences.* Encyclopedia of Earth Sciences Series. Springer, Cham. https://doi.org/10.1007/978-3-030-85040-1_166
30. NOAA. 2004. *Environ. Info. Ser. C-1,* NOAA NCDC, 10p.

Post-discovery on-site measurements

8

On-site measurements are useful in assessing the **accuracy** of all **microclimate** models. They are especially needed for complex mechanistic microclimate models of man-made structures (Chapter 6). This chapter addresses the many questions arising if you decide to make on-site measurements: when, how long, how frequently, and with what instruments.

How long should you measure?

The immediate **climate** around the corpse, the microclimate, changes over minutes and hours (Table 8.1). The influence of large-scale **weather** and **local climate** influences however need to be considered when modeling any outside microclimate. Efforts should be made to get on-site measurements during the types of weather that were reported over the period between the person's last-sighting to corpse discovery (Table 8.1). For example, if this period occurred in the early summer in mid-**latitude** of the US with storm systems rolling through, then the on-site measurements should made during the early summer in similar weather.

If these measurements cannot be made under both similar weather, measurements should be made over a range of weather conditions with **temperature** and **humidity** changes comparable to those during the period of interest. Since large-scale weather systems typically pass over locations in the temperate US climates every week to 3–10 days,[1] measurements should be conducted for at least 10 days to capture the influence of these weather transitions on the microclimate. For climates with longer periods between changes in the weather, the duration of measurements depends on whether the weather transitions occurred in the area over the period between last-sighting and discovery. If corpse location is influenced by local climates (Chapters 4 and 5), measurements should be made for a number of days that include the possible changes in local climate associated with shifts in wind direction or sky condition.

The duration and timing of measurements in **unconditioned** buildings follow the guidelines for measurements made outside. Measurements in **conditioned** buildings should be conducted over several heating and/or cooling

DOI: 10.4324/9781003486633-8

Table 8.1 Scales of climate influence

	Space	Time
Large-scale weather	Miles to 100's of miles	Days to weeks
Local climate	110's of feet to miles	Hours to days
Microclimate	0 to 100's of feet	Minutes to hours

cycles. Measurements should be taken over at least three cycles of heating/cooling to create a relationship between the outside temperature and the number of heating/cooling cycles and the relationship between the thermostat temperature and the air temperature at the discovery location (see Appendix Case 5).

The duration of measurements in the soil or water depends on the depth. The more gradual and diminished response of soil temperatures at 20 in or more depth to weather changes limits the usefulness of extended temperature measurements.[2] Water temperatures near the surface are similar to the air temperature[3] and are influenced by the weather conditions. As with soils, water temperatures vary less with increasing depth[4] If taken, soil or water measurements should made for at least several days for model validation.

If you are using multiple temperature and/or humidity **sensors**, each sensor will respond slightly differently to the environment but usually within the manufacturer specifications. To improve the accuracy of the model for which the measurements are to be used, you should put all the sensors you will use together in an enclosure and determine the difference in sensor response--either before or after the on-site measurement period.

How frequently should you measure?

Statistical analysis (Chapter 7) generally requires measurements to be independent- not depending on a prior value. How quickly a sensor responds to a change in the environment is defined by the **time constant** (T).[5] The T represents the time it takes for the sensor to change 63% of an instantaneous step change in temperature or humidity (Figure 8.1). If we measure every 3T, we have a 95% confidence that the value the sensor is reporting is actually an independent measure of the environment (Figure 8.1).

The T for humidity sensors is commonly less than that for temperature sensors. For temperature sensors, the greater the thermal mass, the longer the T and the less frequent an independent measurement can be taken. Commercially-available combined temperature and humidity loggers with internal sensors typically have T values of about 17 minutes. Commercially-available combined temperature and humidity loggers with external sensors typically have T values of about 4 minutes.

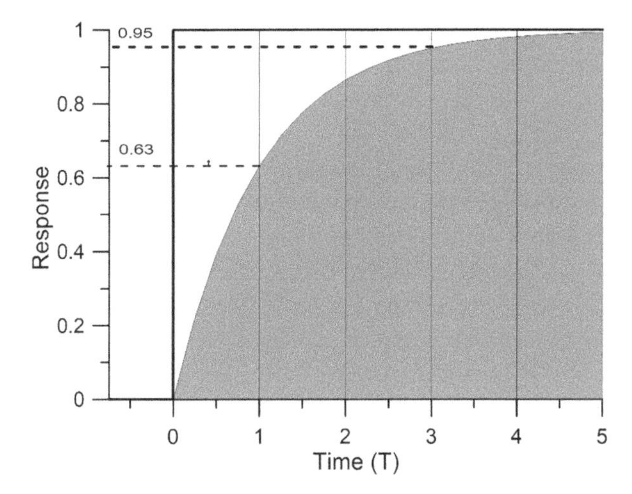

Figure 8.1 The time response of a sensor to a step change in environment. The change in the environment (Black solid line) and the sensor response to the change (grey domain) are indicated.

Measurements of a microclimate should be made more frequently than you need in order to assure that any unusually high or low values are real (repeated) and not an instrument **error**:

- Measurements above-ground outdoors should be **record**ed every 30 minutes[6] using sensors with T less than 10 minutes.
- Measurements below-ground or in water bodies outdoors should be recorded every hour using sensors with T less than 20 minutes.
- Measurements within and around man-made structures should be recorded every 10 minutes[7] using sensors with T less than 3 minutes. Measurements made in conditioned buildings should be made over at least three heating and/or cooling cycles to create a relationship between the outside temperature and the number of heating/cooling cycles during a day.

What equipment do you need?

A sensor system consists of sensor(s), mounting for the sensor(s) including any **radiation** shielding for the sensor, and a **data logger** to record the measurements. Most commercially-available electronic sensors are either internal to the logger and exposed to the air through slits in the housing or external to the logger on a cable.

Important specifications to consider when purchasing/using a system include sensor accuracy and T value and sensor system **resolution**. Sensor system measurement resolutions of 0.1 °**F** for temperature and 1% for **RH** are adequate for all on-site measurements. While manufacturer specification will state the minimum measurement accuracy of the sensor model they produce, you should have your specific sensor accuracy traceable to a standard. Certificates should be available that traces the sensor accuracy to a US National Institute of Science and Technology secondary standard.[8] Such certifications are generally available from the sensor manufacturer at the time of purchase, often for an additional cost.

Traceability of the measurement record is also important. Certification of climate records was discussed in Chapter 7. You are essentially trying to document your on-site measurements similarly. Commercially-available systems can automatically measure at regular intervals, be controlled by your cell phone, notepad or laptop by Bluetooth® technology or to your laptop computer by USB cable and have recorded **data** downloaded with time stamps to these same devices. Hand-held sensor/logger systems should not be used because they introduce the risk of the operator influencing the measurement (e.g. Appendix Case 6).

Temperature measurements

External sensors typically have smaller T values than internal sensors. Data loggers with internal sensors are commercially-available for approximately $100 while those with external sensors cost approximately $200. If the temperature sensor is placed in soil or water, the sensor casing and cable must be water-tight. If the temperature sensor is mounted above-ground outside, it should be shielded from the direct sunlight and surrounding **thermal radiation**. Such radiation shields are either un-ventilated (passive) or ventilated. Generally passive shields are used in on-site measurements because they do not need power. You can either purchase a passive radiation shield from the sensor manufacturer or make your own shield: a wide variety of relatively simple radiation shield designs that have been tested.[9–12]

You should use temperature sensors with accuracies comparable to that of existing climate networks. Standard temperature sensors have accuracies ranging from 0.4 °F to 1.8 °F.[5] The typical error of a Summary-of-Day Climate station **MMTS** sensor with 'Gill' radiation shield is +/− 1.5 °F[13] while that of the temperature component of the hygrometer with 'Gill' shield used in **MesoNet** stations is +/− 0.7 °F.[9] The Palmer Model 35B soil thermometer, used at many Summary-of-Day Climate station has a time response of 4 to 5 minutes and an accuracy of about 4 °F.[14] The soil temperature probe used in most MesoNet stations has an accuracy of +/− 0.4 °F. USGS water temperature measurements on streams, rivers, lakes, and oceans have accuracies of +/− 0.4 °F.[15]

Humidity measurements

Commercially-available sensors are available with internal RH sensors exposed through slits or external sensors on cables. The internal sensor systems should be not be used due to the high T value of these systems. Data loggers with external sensors cost $200 or more. These sensors must be protected from mechanical abrasion and condensing water and consequently cannot be in direct contact with the soil. If this happens, the sensor should be recalibrated. The RH sensor does not need a radiation shield, however a shield is needed if the sensor is also measuring air temperature.

Again, you should use RH sensors with accuracies comparable to that of existing climate networks. Typically, humidity sensors have accuracies ranging from +/– 2% to +/– 15% RH.[5] The **ASOS/AWOS** network RH accuracy is +/– 3%[16] while the RH component error of the typical MesoNet network hygrometer is +/– 3%.[9]

Where should you measure?

The sensor system should be located to maximize the microclimate information attained:

- If the corpse was outside on the ground, temperature or temperature/humidity measurements are needed to describe the microclimate. Sensor(s) need a radiation shield mounted near the height of the corpse.
- If the corpse was partially or wholly buried in some material (soil, water, detritus), temperature sensor(s) should be mounted in that material at the depth of discovery and in the open air above that material. No humidity measurement should be made.
- If the corpse was found in an enclosed space such as a car, crawlspace, trash can, the sensor(s) should be mounted outdoors with a radiation shield near the height of the corpse as well as where the corpse was in the enclosed space.
- If the corpse is discovered in a building, temperature /humidity sensors should be placed where the corpse was discovered and outside the building in a radiation shield. If the building was conditioned, additional temperature measurements should be made at the thermostat (if conditioning was active) and at the heating/cooling vent in the room where the corpse was discovered.

References

1. Dickson, R.R., 1971. *Journal of Applied Meteorology and Climatology*, 10(2), pp.186–193.
2. Carson, J.E., 1961. *ANL-6470*, Argonne National Lab., IL, USA, 244p.
3. Edinger, J.E., Duttweiler, D.W. and Geyer, J.C., 1968. *Water Resources Research*, *4*(5), pp.1137–1143.
4. Stefan, H.G. and Preud'homme, E.B., 1993. *Journal of the American Water Resources Association*, *29*(1), pp.27–45.
5. DeFelice, T.P., 1998. *An introduction to meteorological instrumentation and measurement*. Upper Saddle River, NJ: Prentice Hall. 229p.
6. Gough, W.A., Žaknić-Ćatović, A. and Zajch, A., 2020. *International Journal of Climatology*, *40*(13), pp.5451–5463.
7. Wang, Z., 2022. *Building and Environment*, *222*, p.109464.
8. Bruce, S., Possolo, A. and Watters, R., 2021. Metrological Traceability Frequently Asked Questions and **NIST** Policy, Technical Note (NIST TN), National Institute of Standards and Technology, Gaithersburg, MD, [online], https://doi.org/10.6028/NIST.TN.2156, https://tsapps.nist.gov/publication/get_pdf.cfm?pub_id=932360 (Accessed February 29, 2024)
9. Erell, E., Leal, V. and Maldonado, E., 2005. *Boundary-layer Meteorology*, *114*, pp.205–231.
10. Tarara, J.M. and Hoheisel, G.A., 2007. *HortScience*, *42*(6), pp.1372–1379.
11. Navarro-Serrano, F., López-Moreno, J.I., Azorin-Molina, C., Buisán, S., Domínguez-Castro, F., Sanmiguel-Vallelado, A., Alonso-González, E. and Khorchani, M., 2019. *Atmospheric Research*, *224*, pp.168–179.
12. Maclean, I.M., Duffy, J.P., Haesen, S., Govaert, S., De Frenne, P., Vanneste, T., Lenoir, J., Lembrechts, J.J., Rhodes, M.W. and Van Meerbeek, K., 2021. *Methods in Ecology and Evolution*, *12*(8), pp.1397–1410.
13. Hubbard, K.G.; Lin, X. and Walter-Shea, E. 2001. *J. Atmos. Oceanic Technol.* 18:851–864.
14. NWS, 1989. *Cooperative Station Observations. National Weather Service Observing Handbook No. 2*, NWS NOAA, Silver Springs, MD, USA, 83p.
15. Wagner, R.J., Boulger Jr, R.W., Oblinger, C.J. and Smith, B.A., 2006. *United States Geological Survey Techniques and Methods 1-D3*, 51p.
16. Sun, B., Baker, C.B., Karl, T.R. and Gifford, M.D., 2005. *Journal of Atmospheric and Oceanic Technology*, *22*(6), pp.679–686.

Modeling the corpse microclimate

9

The estimation of the daily average air **temperature**, air **humidity**, soil temperature, or water temperature at the corpse location can be approached a number of different ways. The estimation of the daily values of these measures can be approached using direct, empirical, theoretical, and mechanistic approaches. Here we seek to give broad-brush descriptions of these approaches. The applicability of a given modeling approach in the estimation of the **microclimate** depends on the:

- availability of ancillary information for the corpse location (partially listed in Chapter 6),
- proximity, frequency, and quality of available measurements (Chapter 7), and
- complexity of the microclimate to be modeled (Chapter 6).

Estimating the air temperature

Direct spatial extrapolation of the conditions at the discovery location from the nearby **climate** station is often used in **PMI** determinations. Daily air temperatures at a climate station and a discovery location can be anticipated to be similar for separation distances of between 1 and 30 miles provided they both have similar microclimates.[1, 2] Methodologies are discussed in Chapter 7.

This approach assumes that the microclimates of both locations are very similar and the temporal range of conditions where the corpse is found are well-approximated by a nearby climate station measurement **record**. Problems with the indiscriminate use of this direct approach have been well documented.[3, 4] The direct use of a climate record at a corpse location was used in Appendix Cases 1 and 6.

Usually, adjustments to the climate record must be made to account for differences in the microclimates of the measurement and corpse locations. This can be done either empirically or mechanistically.

An empirical relationship between the air temperature records from a long-term climate station and short-term on-site measurements at the

DOI: 10.4324/9781003486633-9

discovery location can be developed if the microclimates of the two locations respond similarly to long-term **weather** conditions and **local climate** influences. The relationship is then used to estimate the air temperature at the location of discovery. This approach consequently involves the collection of post-discovery on-site measurements discussed in Chapter 8.

The more similar the local climates of the corpse location and climate station are, the more likely such an empirical approach will work. Of the linear models predicting the air temperature of a mountain location, a farm, a rooftop, and in two closed spaces based on air temperature of a nearby climate station, the greatest **accuracy** was for the farm location and least for the mountain location.[5] In another study, the **errors** of such empirical models where so large that they did not improve the estimate of average 5-day, 10-day or 15-day air temperatures over grass, fallow fields, under a vegetation canopy, near a river, and in a number of unheated shelters based on climate stations within 9 miles distant.[6]

However, the lack of physical explanation for why air temperatures at the two locations appear to be related for one period in time makes extrapolation of that relationship to a different time period questionable. For example, linear correlations for short periods of time within a 300-day period may differ substantially from the overall relationship. Consider two measurement locations: one in a field in the mountains and one at an airport on the plain near the mountains. A general linear correlation relationship is evident for air temperatures with variations related to differences in local climate influences and large-scale weather systems. (Figure 9.1). However, the relationship does not hold for all 10-day intervals (Figure 9.1). We conclude that this method can be used to fill in missing values across a period of time with similar weather conditions at both locations (Chapter 7) but should not be used to predict temperatures during periods of different weather conditions (such as cold **front** passages) and different seasons.

Mechanistic models of the microclimate at the corpse location are constructed from the understanding of the physical processes influencing the **sensible heat** energy budget (Chapter 4). A complete energy budget model includes heat transfers by **radiation**, **convection**, **advection**, **conduction**, and **evaporation** (Chapter 4) and would require canopy cover, slope and aspect information as well as hourly measurement records of wind speed and direction, humidity, **cloud cover**, air temperature, and **precipitation** or surface and atmospheric moisture content.[7, 8] These models can be applied without regard to differing weather conditions or seasons.

The typical mechanistic model used for a PMI estimation is based on various assumptions and simplifications balancing accuracy and availability of information and accuracy of the model. A common simplification of the energy budget is to assume the conduction of heat into the ground and the

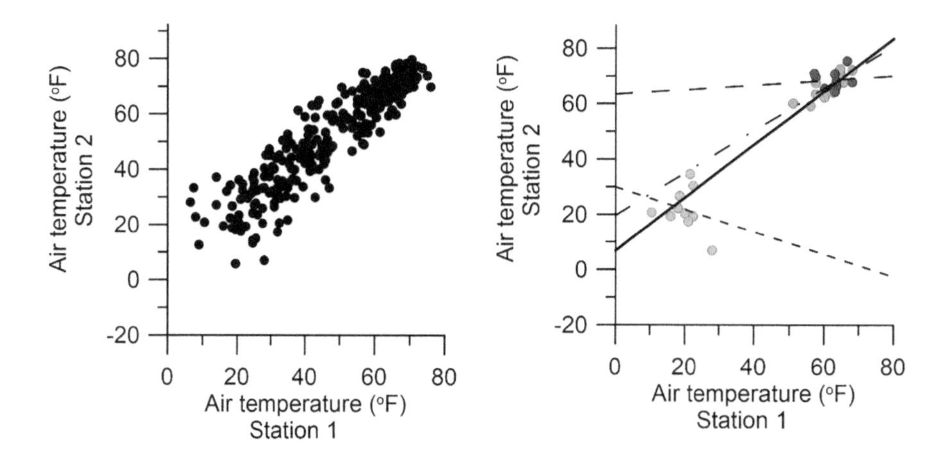

Figure 9.1 Empirical relationships of daily air temperature between two stations. The relationship of daily air temperatures between two stations over 300 days (left) and over three 10-day intervals during different portions of the period (right). The linear regressions between the two stations air temperatures are indicated with different line patterns. The linear correlation of the two stations for the entire 300-day period (solid line) and the correlations of the same stations for 3 10-day periods within the 300 days (dashed lines) are indicated.

Data source: NOAA National Center for Environmental Information, https://www.ncei.noaa.gov/access

evaporation of heat into the air is small.[7] No thermal advection is often assumed but can usually be modeled. Other simplifications of the energy budget are based on the corpse location:

- If the corpse is located outside under a vegetation canopy (Chapter 6), the air temperature can be modeled assuming the average temperature above a vegetation canopy and in **open terrain** are equal and the heat transfer from the canopy top to the floor is through convection. If hourly climate information is available, the air temperature can be estimated using a simplified energy budget model including exchange terms discussed in Chapter 6 for the location under the canopy.[9]
- If the corpse is outside located near a terrain transition (Chapter 6), the advection of heat needs to be modeled. Since the effect of a terrain transition on air temperatures varies with wind direction, wind-direction segregated measurements from a nearby hourly climate station can be used to estimate the transition effect.
- If the corpse is found in an enclosure (Chapter 6), we often assume all walls/ceiling/floor inside the enclosed volume are at a constant temperature over the course of a day corresponding to the various

material's emissivity and **heat content.**[10, 11] Complete energy budget models[12, 13] with hourly inputs should be used to model the air temperature outside the enclosure as well as transfer of energy to the inside.

The accuracy of any mechanistic model is typically improved when on-site measurements are made to account for specific microclimate differences between the location of discovery and the climate station. Examples of using simplified mechanistic approaches are found in
Appendix Cases 2, 3, and 8.

Estimating the soil temperature

Soil temperatures vary more gradually over time than air temperatures and often not measured at a climate measurement station. Since the soil temperatures are strongly dependent on the soil conductive properties and soil moisture content, even if measurements are made at a climate station they likely do not apply to the soils found at the corpse location (Chapter 6). Even rarer than measuring soil temperatures is the measurement of soil temperature deeper than 4 inches. The modeling of soil temperature at the depth of corpse burial is difficult since there are rarely measurements nearby to reference.

The most common approach is to theoretically estimate the soil temperature from the measured or modeled air temperature by assuming: 1) the air temperature follows a wave-like pattern over the day (Figure 4.2) and year and 2) the range in soil temperatures diminish with depth depending on the dry soil properties.[10, 14] More realistic modeling of the hourly air temperature over the day can increase the model accuracy.[15]

Soil temperatures can also be mechanistically modeled using a sensible heat budget model to determine the temperature of the soil surface and theoretically model the heat transfer into the soil. This requires an hourly air temperature, wind speed, **relative humidity**, cloud cover as well as information on soil thermal properties and soil wetness.[16] The disturbance of the soil due to the corpse burial will change the soil properties and require some estimation from literature.

Estimating the air humidity

Provided there is no local climate influence (Chapter 5), the **dew point temperature** or **specific humidity** does not vary as much as air temperature over distance. Assuming the humidity of the air does not change, the relative humidity at a location is directly determined from the measured dew point temperature and the modeled air temperature:

- If the dew point temperature is greater than the modeled air temperature, the relative humidity is calculated according to Campbell and Norman.[10] If the dew point temperature is less than or equal to the modeled air temperature, the relative humidity is 100%.
- If the air temperature is below 32 °F, the relative humidity depends on whether the water condensed from the air is in liquid or solid form (See Chapter 5). An example of using this approach is found in Appendix Case 8.
- If advection of humidity is likely (near the ocean, a lake, or river), this process should be evaluated. Since the influence of a water body varies with wind direction, wind-direction segregated measurements from a nearby hourly climate station can be used to estimate the influence. Adjustments to the daily climate records can then be made using the estimated lake effect each day. If water temperatures, air temperatures, cloud cover, wind speed and direction have been measured, moisture advection can be directly computed.[17]
- If there are no hourly climate station records with wind and humidity measures available, an approximation of the advected moisture can be directly made using documented effects from similar sized lake and distances from the lake as discussed in Chapters 5 and 6. If hourly climate records are available, a mechanistic approach can be taken to estimate the advected moisture from the water body using hourly wind speed and direction in combination measurements of humidity at a climate station. With this information the relative influence of the advection can be estimated based on the differences in distance from the lake to the climate station and the discovery location.
- If the corpse is found under vegetation canopies, the influence of the canopy depends on the density of the canopy. If the canopy is relatively dense, the relative humidity under the canopy can be assumed to be approximated by that reported in the literature (Chapter 6).
- If the corpse is found in an enclosed space or building (Chapter 6) the relative humidity of can be estimated either assuming the water content of the outside air and the equilibrium air temperature of the space or by applying adjustments based on literature values discussed in Chapter 6. Since the air from outside a building typically replaces the volume of air inside a building in two to ten hours (Chapter 6), the daily average relative humidity within buildings can be estimated assuming the water content of the outside air and the equilibrium air temperature of the space. On-site post-discovery air temperature measurements will greatly aid in the modeling effort.

Estimating the water temperature

Water temperatures vary more gradually over time than soil or air temperatures. However, lakes can overturn multiple times a year-- creating sudden changes in water temperatures at a given depth. Broad assumptions are usually needed to estimate the water temperature over extended periods of time at the specific location of corpse discovery or estimated location of the corpse on a given day.

If water temperatures for the water body are available, the water temperature at the corpse location can be estimated using the distance between corpse location and measurement and rate of change in temperature with distance for similarly-sized rivers documented in the literature (Chapter 6). Since water temperatures are generally made at depth while corpses are generally found at or near the surface, a theoretical or mechanistic model is needed to estimate the water temperature at the corpse depth.

The most common approach to modeling water temperature at depth is done theoretically assuming: 1) good mixing of the water by the wind, 2) that the water temperature is in equilibrium with the air temperature,[18] and 3) the temperature varies in a wave-like pattern over the day and year.[19] The rate of change of the average water temperature with depth is often assumed to be a fixed value within the well-mixed layer in lakes.[19] Linear correlations between the air temperature and water temperature can be made using on-site measurements.[20]

The water temperature of small streams is more difficult to estimate than lakes or large rivers since typically there are few to no measurements and the streambed can be heated by the sun which then heats the water. These streams can be mechanistically modeled using estimates of the spatially distributed canopy cover, streamflow, stream characteristics (Chapter 6) and the hourly local weather conditions across the watershed.[9] Post-discovery measurements of the water and air temperature at the discovery location will aid in the modeling effort. Barring this, the initial headwater water temperature can be approximated by a literature-derived groundwater temperature and rates of cooling or heating estimated from **sky view** (Chapter 6).

Statistics describing model accuracy

Regardless of the approach used to estimate the corpse location microclimate, an assessment of the error in the model is needed. If the microclimate estimation is based on literature-based 'adjustments' to the climate station record, the error of the estimations can only be assumed. If the corpse location microclimate estimation is based on theoretical or mechanistic modeling, the error assessment is best accomplished using post-discovery on-site measurements and is generally limited to period(s) of post-discovery measurements. Unfortunately, the same post-discovery measurements are also usually used to 'adjust' any model for any un-accounted-for local conditions. Consequently, the estimation of error actually only represents the unexplained daily differences between the model and the measurement during that specific period of measurement and may or may not represent the error in applying the model to the entire period from last sighting to corpse discovery.

Modeling errors are generally described using the '**correlation coefficient**', the 'root mean squared error' (**RMSE**), the 'mean **bias** error' (**MBE**), and/ or the 'mean absolute error' (**MAE**). As with all statistics, these statistical measures assume the validity of assumptions about the frequency of occurrence of the magnitudes of daily values being measured and modeled. The frequency of occurrence of values over a period of time describes the '**frequency distribution**' of the measurement and is determined for intervals of values of the measurement- for instance the frequency that the daily air temperature was between 40 °F and 50 °F during the period of interest. When values are symmetrically distributed above and below a most-common central value (most often occurring over the period of time) and the distribution is called a '**normal distribution**'.[21]

Frequency distributions of daily air temperature

Daily maximum and minimum air temperatures over an entire year often have an annual temperature periodicity. The annual wave pattern in the minimum and maximum daily air temperatures results in temperatures that do not vary uniformly around an average value- resulting in a non-normal frequency distribution (Figure 9.2; panels A and B). The daily air temperatures determined from averaging these two temperatures are also not normally distributed. Non-normal distributions occur due to annual changes in sunlight intensity, changes associated with large-scale weather patterns, and local climate influences that shift the frequency of occurrence of a given daily air temperature over days and weeks. Daily air temperatures at some locations with steady winds and cloud cover may be normally distributed over weeks within some seasons and climates.[22]

Frequency distributions of daily air humidity

Atmospheric humidity, as previously discussed, need to be first represented as a water content of the atmosphere (a specific humidity or dew point temperature) before we can determine an average **RH**. The range of daily average humidity values tend to approximate a normal distribution, although there are typically a higher frequency of dewpoints above the average values than

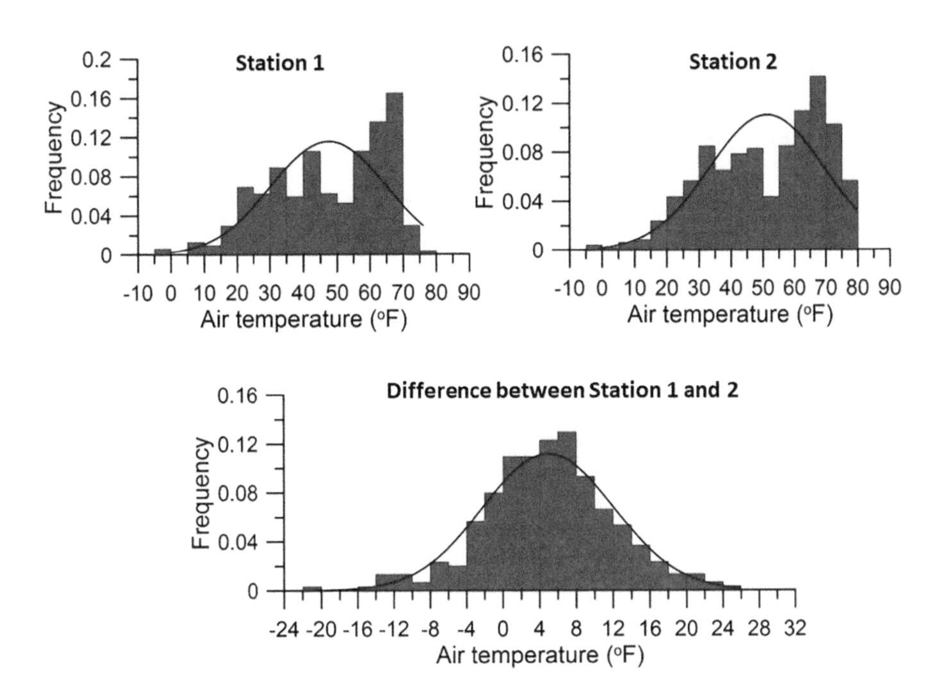

Figure 9.2 Relationships of air temperature distributions at two stations over 300 days. Relationships of air temperature frequency distributions at two stations. The daily air temperature distribution for two stations and the distribution of the difference in temperatures between the two stations is illustrated. Expected variation in air temperatures if they were normally distributed are indicated with solid lines. The distributions for the two stations do not fit the solid line and are not normally distributed. The difference between the two temperature measurements does fit the normal distribution line and can be considered normally distributed.

Data source: NOAA National Center for Environmental Information https://www.ncei.noaa.gov/access

below the average value since the amount of water the air can hold increases rapidly with increasing air temperatures[22] (Figure 5.1).

The specific humidity and dew point temperature of the air tends to change more gradually across space (horizontally and vertically) and over time than the air temperature (see Chapter 5). Hubbard[1] found similar humidity values within a distance of about 18 mi across the relatively uniform terrain of the high plains of the USA.

While the dew point or specific humidity of the air does not change quickly, RH changes as air temperature changes (Figure 5.1). So, humidity model accuracy should be based on the estimated dew point temperature or specific humidity and not RH. The distributions of these measures of humidity at a location are generally near-normal over months and years (Figure 9.3). Large changes in dew point temperature can occur with the passage of weather

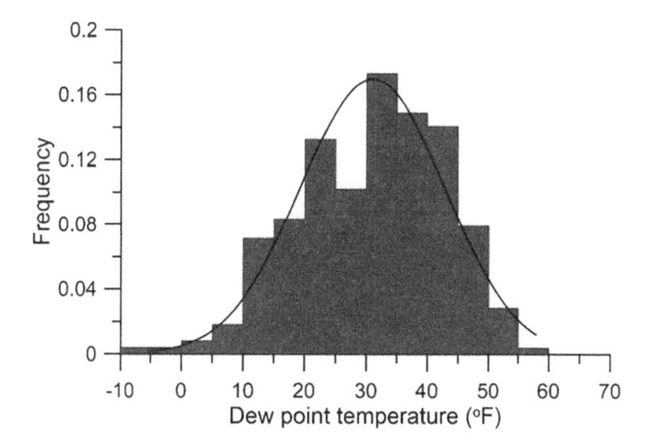

Figure 9.3 Frequency distribution of daily average humidity values at one station over 300 days. A normal distribution about the average value is indicated with a solid line.

Data source: NOAA National Center for Environmental Information https://www.ncei.noaa.gov/access

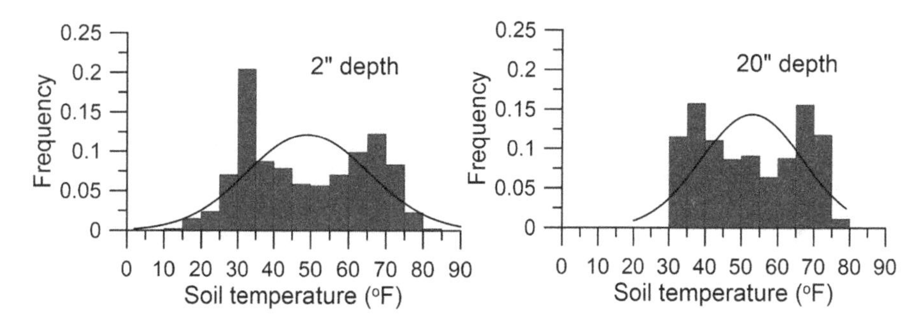

Figure 9.4 Frequency distributions of daily average soil temperatures at two depths at one station over 11 years in Indiana. A normal distribution about the average value is indicated with a solid line.

Data source: Indiana State Climate Office, W. Lafayette, IN

systems can result in distinctly non-normal dew point temperature frequency distributions over short periods of time such as weeks.

Frequency distributions of daily soil temperatures

Soil temperatures are also not normally distributed. The frequency of occurrence of a given daily average soil temperature will have two frequency distribution peaks[23] because the soil temperature varies annually in a wave pattern similar to the air temperature. The magnitude of the peaks decreases with increasing depth in the soil (Figure 9.4). If the soil freezes, there will typically be high frequency of temperatures at freezing (Figure 9.4).

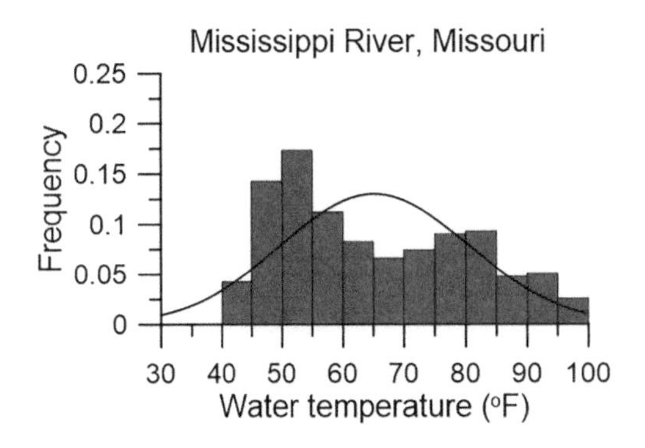

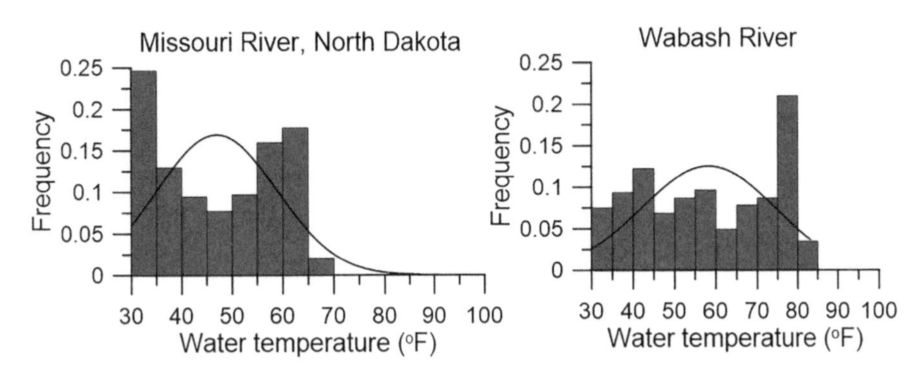

Figure 9.5 **Frequency distributions of daily average water temperatures of various river locations over four years**. A normal distribution about the average value is indicated with a solid line. The annual variation in daily temperatures of rivers with a range of width and flow volumes are indicated (Mississippi River> Missouri River > Wabash River).

Data source: USGS:water.usgs.gov. URL: https://waterdata.usgs.gov/ca/nwis?

Frequency distributions of daily water temperatures

Water temperatures are also not normally distributed. The frequency of occurrence of a given water temperatures in lakes and streams has an annual wave pattern of average daily water temperatures near the surface. If the location is in a temperature or cold climate where the water freezes during part of the year, the frequency of temperatures at freezing is greatly increased (Missouri River, Figure 9.5). Low water levels can also shift the water temperature distribution since the **sensor** is at a fixed position and approaches the surface temperature when the river is low (Wabash River, Figure 9.5).

Statistical measures of model error

Linear model accuracy is often assessed using a correlation coefficient or the squared value of this coefficient called the '**coefficient of determination**'.

These statistics assume that the distributions of both the modeled and measured records are normal- which is generally not true for daily air, soil, or water temperature. However, the distribution of the difference between two values (modeled and measured or on-site and climate station) of daily air, soil, or water temperatures tends to be normally distributed[24] (Figure 9.2 panel C). This gives us the ability to use MAE, MBE, and RMSE statistics to describe modeling errors.

The RMSE statistic describes the variability in differences between the modeled and measured values. This statistic is the best single descriptor of model error.[25, 26] However, the statistic is strongly influenced by a few large model errors and thus can mis-represent the model accuracy.[27]

Modeled microclimates frequently follow the daily pattern of measured conditions but tend to be higher or lower than post-discovery on-site measured values. This tendency is called a 'bias'. While the MBE statistic is intended to give an overall representation of the model over- or under-prediction, it may under-represent the overall model error.[27] The MAE statistic describes the absolute value of the bias and is considered to be the best single measure of bias error.[27]

Issues in microclimate modeling

An accurate model of the air temperatures and humidity at the location of corpse discovery is desired but is always limited by the available information. Direct extrapolation of climate measurements is easy to accomplish but rarely a good approach. Similarly correlating the conditions at two locations without consideration of differences in microclimates will generally provide a poor estimation of the corpse location microclimate. It is important to address differences in microclimates of the climate station and the corpse location either through literature-based adjustments to the climate record or mechanistically-based modeling.

Regardless of the approach used to estimate the microclimate at the corpse location, an assessment of the error in the model is necessary. Post-discovery on-site measurements always improves the estimate of the conditions at the corpse location- even if only to provide an error estimate.

References

1. Hubbard, K.G., 1994. *Agricultural and Forest Meteorology*, *68*(1–2), pp.29–41.
2. Daly, C., 2006. *International Journal of Climatology: A Journal of the Royal Meteorological Society*, *26*(6), pp.707–721.
3. Dabbs, G.R., 2010. *Forensic Science International*, *202*(1-3), pp.e49–e52.
4. Giles, S.B., Harrison, K., Errickson, D. and Márquez-Grant, N., 2020. *Forensic Science International*, *315*, 110419.

5. Jeong, S.J., Park, S.H., Park, J.E., Park, S.H., Moon, T.Y., Shin, S.E. and Lee, J.W., 2020. *Forensic Science International, 309*, 110196.
6. Dourel, L., Pasquerault, T., Gaudry, E. and Vincent, B., 2010. *Psyche: A Journal of Entomology, 2010*, 610639.
7. Bennie, J., Huntley, B., Wiltshire, A., Hill, M.O. and Baxter, R., 2008. *Ecological Modelling, 216*(1), pp.47–59.
8. Maclean, I.M., Duffy, J.P., Haesen, S., Govaert, S., De Frenne, P., Vanneste, T., Lenoir, J., Lembrechts, J.J., Rhodes, M.W. and Van Meerbeek, K., 2021. *Methods in Ecology and Evolution, 12*(8), pp.1397–1410.
9. Lee, R., 1978. *Forest microclimatology.* Columbia University Press, New York, 276p.
10. Campbell, G.S. and Norman, J.M. 1988. *An introduction to environmental biophysics.* 2nd Ed., Springer-Verlag, NY, NY. 286p.
11. Janna, W.S., 2000. *Engineering heat transfer.* 2nd Ed., CRC Press, 683p.
12. Oke, T.R., 1978. *Boundary Layer Climates.* Methuen and Co., NY, NY, 372p.
13. Rutkowski, H., 2002. *Manual J Residential load calculations*, 8th Ed., Air Conditioning Contractors of America, Arlington, VA.
14. Prangnell, J. and McGowan, G., 2009. *Forensic Science International, 191*(1–3), pp.104–109.
15. Lei, S., Daniels, J.L., Bian, Z. and Wainaina, N., 2011. *Environmental Earth Sciences, 62*, pp.1123–1130.
16. Kearney, M.R., Shamakhy, A., Tingley, R., Karoly, D.J., Hoffmann, A.A., Briggs, P.R. and Porter, W.P., 2014. *Methods in Ecology and Evolution*, 5(3), pp.273–286.
17. Atkinson, B.W., 1981. *Meso-scale atmospheric circulations.* Academic Press, London, 495p.
18. McCombie, A.M., 1959. *Limnology and Oceanography*, 4(3), pp.252–258.
19. Edinger, J.E., Duttweiler, D.W. and Geyer, J.C., 1968. *Water Resources Research*, 4(5), pp.1137–1143.
20. Stefan, H.G. and Preud'homme, E.B., 1993. *Journal of the American Water Resources Association, 29*(1), pp.27–45.
21. Harmel, R.D., Richardson, C.W., Hanson, C.L. and Johnson, G.L., 2002. *Journal of Applied Meteorology and Climatology, 41*(7), pp.744–753.
22. Brooks, C.E.P. and Carruthers, N., 1953. *Handbook of statistical methods in meteorology.* Air Ministry, Meteorological Office, M.O. 538, Her Majesty Stationary Office, London. 412p.
23. Toth, Z. and Szentimrey, T., 1990. *Journal of Climate*, 3(1), pp.128–136.
24. Kale, B.K., 1970. *The American Mathematical Monthly, 77*(9), pp.992–995.
25. Chai, T. and Draxler, R.R., 2014. *Geoscientific Model Development, 7*(3), pp.1247–1250.
26. Hodson, T.O., 2022. *Geoscientific Model Development, 15*(14), pp.5481–5487.
27. Willmott, C.J. and Matsuura, K., 2005. *Climate Research, 30*(1), pp.79–82.

How do measurement and modeling errors affect the specificity of PMI estimates

10

Accumulated heat units or average air **temperature** and **RH** are typically used in conjunction with either insect development or decomposition evidence to determine **PMI**. These **microclimate** conditions are useful in PMI determinations primarily over time intervals of between 24 hours and six months. Thirty percent of US homicide PMIs were between two days and two weeks, with an additional 23% beyond two weeks.[1] The average PMI in rural Finland is about 5 ½ days.[1] As might be expected, the PMI is typically longer when the corpse is concealed: 42% of concealed homicides are discovered between a week and a year after the homicide.[2]

Some PMI estimation models use averaged air temperature and **humidity** and measures of corpse decomposition,[3] while others use **ADD** and measures of corpse decomposition[4, 5] (Chapter 3). PMI determination based on corpse decomposition is enhanced by hot conditions and inhibited by freezing conditions (Chapter 3). As a result, the loss of all soft tissue can take anywhere from two weeks to six months.[6] Put another way, it is estimated that an ADD (base 32 °F) of 2313 °F days +/- 198 °F days is needed for complete soft tissue decomposition.[3] Since low humidity also inhibits decomposition, estimates of the humidity around the corpse are also included in some estimates of PMI.

Other PMI determinations are based on insect larvae development. Since the development of forensically-important insect larvae depend not on time but on available heat (Chapter 2), the time period for a certain larval stage can be compressed during hot periods and extended in cold periods. Since larvae development typically does not occur until some threshold temperature is reached (called the **base temperature**), the daily or hourly air temperatures used to represent the heat available for development per day or hour are decreased by this base temperature before being accumulated. So, the incremental ADD for a day is zero if the daily average air temperature is less than the base temperature, and the incremental **ADH** for an hour is zero if the hourly average temperature is less than the base temperature.

DOI: 10.4324/9781003486633-10

Approaches to modeling the **microclimate** at the corpse to determine the air temperature and RH vary depending on the availability of **climate** information, the proximity of that information, and the differences in the climate station and corpse location environment (Chapters 6 and 9). Historically, assessments of PMI using either air temperature, humidity, ADD or ADH have often been determined assuming the microclimate at the corpse is identical to that of the closest climate station. But corpses are usually not found in open grassy areas- the environment of most climate stations and the climate station- and often have different **local climate** influences of topography (hills, lakes, streams). In addition, the different environments may be influenced differently by wind, rain, snow resulting from the passage of large-scale **weather** systems (Chapters 4, 5).

Differences between ADD calculated from modeled microclimate temperatures or the direct use of the nearest climate station vary with location and season. Examples of the difference in ADD estimates between assuming the conditions at the nearest climate station versus modeled microclimate at the corpse location include: 71 °F day over a 10-day period in an enclosed storage unit in the fall (Appendix Case 5) and 21 °F day over a 20-day period in an unheated building in spring (Appendix Case 3).

But remember that the estimated air temperature and humidity around the corpse is just an estimate. Each estimate has an **error bound** (estimated value + or – the **error**) in which there is a 95% probability that the actual true value can be expected. And this error bound includes errors due to the measurement (Chapter 7) and errors due to modeling the microclimate (Chapter 9). Random microclimate model errors (represented as **RMSE**) need to be incorporated into the error bounds of the averaged air temperature and humidity or accumulated heat. There will likely also be **bias** errors (represented as MBE) in the modeled air temperature or humidity conditions. The MBE can be simply added or subtracted from the measured value, shifting the estimated value but not changing the error bound around the estimated value. Modeling errors are added to the measurement errors prior to accumulation or averaging over time. For purposes of illustration, we will discuss below the propagation of errors we will address the propagation of measurement errors only – assuming no modeling errors.

Errors in PMI based on air temperature and humidity averages

The error of an average of individual measurements is less than the error of the individual measurements being averaged.[7] Therefore, the longer the period of time that the hourly or daily estimates are averaged, the smaller the error of

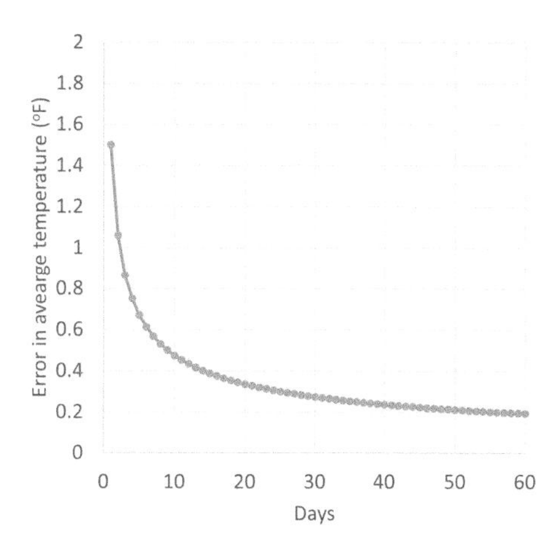

Figure 10.1 Improvement of air temperature average error over time. Improvement in average daily temperature error of measurements made at a 'Summary of the Day' station due to increasing number of values being averaged. It is assumed here that there is no modeling error.

the average (Figure 10.1). For instance, an individual day's air temperature measurement error at a 'Summary of the Day' station, based on the average of daily minimum and maximum temperatures, is +/− 1.06 °F (Chapter 7). For a 14-day average air temperature, this error reduces to +/− 0.4 °F because there are 14 daily temperature values being averaged (Figure 10.1).

The daily average air temperature, calculated from the 24 hourly **ASOS/AWOS** and **MesoNet** station **records**, has an error of +/− 0.35 °F and +/− 0.14 °F respectively. These errors are significantly less than the corresponding error of the 'Summary of the Day' station above. The error for a two-week average air temperature is 0.09 °F for the ASOS/AWOS stations and 0.04 °F for the MesoNet stations.

Humidity measurements are made at ASOS/AWOS and MesoNet stations (and many other automated stations), but not at 'Summary of the Day' stations (Chapter 7). As with averaging air temperatures, averaging humidity values over time decreases the measurement error (Figure 10.2). However, RH should not be averaged since the RH depends on both how much **water vapor** is in the air and how much water vapor the air can hold without droplets forming in the air (Chapter 5). Determining the average humidity over a day from hourly measurements first requires the conversion of measured hourly RH or **dew point temperatures** to specific humidity, then the averaging of air temperature and averaged dew point temperature or **specific humidity**, then calculating a RH based on the average air temperature and dew point temperature or specific humidity. The error in estimating RH from

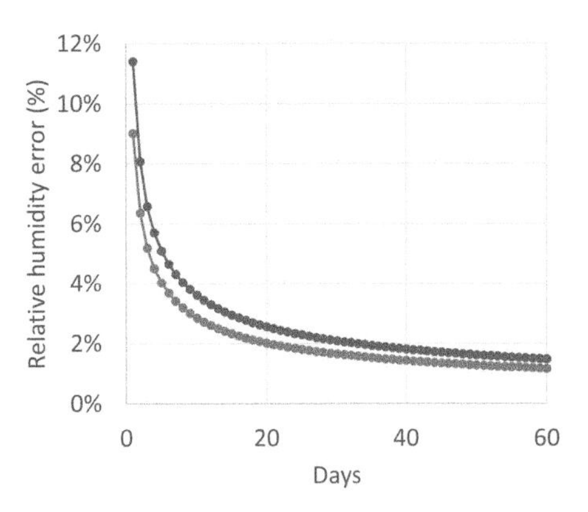

Figure 10.2 Improvement in average daily relative humidity error over time. The average daily relative humidity error estimate based on air temperature and dew point measurement errors of ASOS/AWOS stations (orange circle and line) or air temperature and relative humidity measurement errors of typical MesoNet stations (blue circle and line) due to increasing number of daily mean values being averaged. It is assumed here that there is no modeling error.

dew point and air temperatures at ASOS/AWOS stations (+/− 9%) is greater than that using air relative humidity and air temperature measurements at a nominal MesoNet station(+/− 7%). However, the difference in RH error between these two types of stations decreases rapidly in time such that they differ by less than 1% in four days (Figure 10.2).

We can conclude three things from this: one, it is better to use air temperature information from ASOS/AWOS or MesoNet stations than 'Summary of the Day' stations, two there is little difference in average relative humidity errors associated with ASOS/AWOS or MesoNet stations, and three the shorter the PMI, the greater the error in estimated average daily air temperature or humidity.

The overall PMI error bound of a PMI based on the combined errors of average air temperature and RH in combination with corpse decomposition state[3] should be based on the four possible extreme microclimate scenarios:

- upper bound (estimate + error) for both air temperature and RH,
- lower bound (estimate − error) for air temperature and RH,
- upper bound for air temperature and lower bound for RH,
- lower bound for air temperature and upper bound for RH.

The minimum PMI will be when the lowest value of the four error combinations.

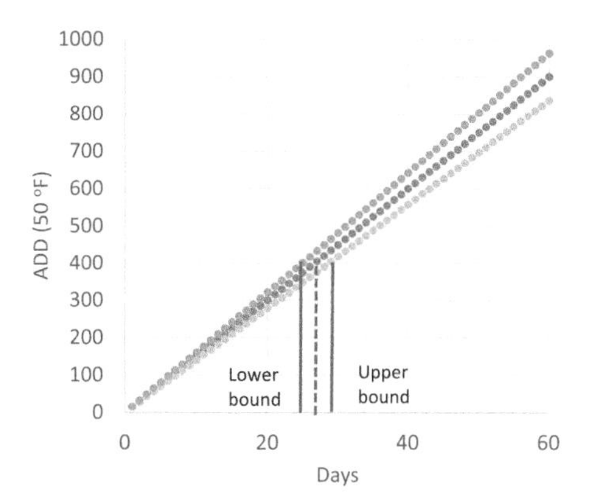

Figure 10.3 Decreasing accuracy of Accumulated Degree Day estimates (base 50 °F) over time. The ADD based on 'Summary of the Day' measurement errors with increasing time interval from day of body discovery assuming an average daily temperature of 65 °F are indicated for 60 days. The average ADD (orange circle) and upper and lower bounds of the error in ADD (yellow and green circles respectively) are indicated. The corresponding day when 400 ADD is reached is indicated by the dashed line. The range in number of days in which 400 ADD might be attained is indicate by the solid vertical lines. It is assumed here that there is no modeling error.

Errors in PMI based on daily heat accumulations

While the error in average daily air temperature and RH decreases as the time period being averaged increases, the ADD error increases with the time period accumulated (Figure 10.3). The ADD error bound guides the entomologist and anthropologist in estimating the probable range in PMI. So, days with higher average daily air temperatures result in fewer days needed to reach a given estimated ADD and smaller error bound (in days) for that estimate.

The impact of relatively small changes in daily ADD increments on PMI can be large depending on the season: there was a 20 °F day reduction in ADD after modeling the under-canopy microclimate of a forest during the winter that corresponded to a three-day shift in the estimated PMI (Appendix Case 1).

Since averaging values decreases the error of the averaged value, there are significant advantages to using climate stations that report hourly measurements rather than daily measurements to estimate ADD. The error in average daily air temperature measurements

- based on a daily maximum and minimum temperature made at all 'Summary of the Day' climate stations (Chapter 7) results in an ADD error of +/– 1.06 °F day.
- based on 24 hourly measurements made at ASOS/AWOS and MesoNet climate stations results in an ADD error of +/– 0.35 °F day for ASOS/AWOS stations and +/– 0.14 °F day for typical MesoNet stations.

Reducing the ADD error bound narrows the range of days needed to reach that ADD. The error range in days needed for a 210 ADD (base 50 °F) value is less than one day for both ASOS/AWOS and MesoNet stations compared with a PMI bound between 7 days earlier to 12 days for the' Summary of the Day' stations assuming a constant daily air temperature of 55 °F. As the average daily temperature increases, the advantage of calculating daily averages from hourly measurements instead of from maximum and minimum temperature measurements decreases: the PMI error for an ADD of 210 °F day in a climate with constant daily air temperatures of 75 °F is +/– 1 day for 'Summary of the Day', MesoNet, and ASOS/AWOS stations.

Case examples of estimated error-bounded ranges in ADD (base 50 °F) between a person's last sighting to their corpse discovery include 42 °F days for a case in Texas for a 22-day period in the winter (Appendix Case 1), 32 °F days for a corpse in Tennessee for 10-day period in the Fall (Appendix Case 5) and 53 °F days for a corpse in New Mexico for 40-day period in the winter (Appendix Case 6). Extended periods of time between a person's last-sighting and corpse discovery result in large error bounds in ADD (base 50 °F): 1100 °F days for a corpse over 252 days in the mountains in Montana (Appendix Case 8).

When the time between last-sighting and corpse discovery is less than a week, it is preferable to estimate the heat available for insect development and decomposition using hourly values. In this case, ASOS/AWOS, MesoNet or other hourly (or more frequently) reporting stations must be used to estimate the ADH. Since there are typically air temperature variations over the day, the ADH values undulate (Figure 10.4). Since air temperatures vary annually in **temperate climates**, the number of days it takes and the error range in days it takes for a given ADH or ADD will vary over the year.

If we assume measurements are made at an ASOS/AWOS station and the daily range in air temperatures is 20 °F, the accumulation of ADH and the corresponding error in the estimated day for that ADH to be accumulated varies by season. For a daily air temperature of 75 °F (resulting in air temperatures always greater than the 50 °F base temperature), an ADH (base 50 °F) of 1200 °F hour (corresponding to 50 °F day) occurs after only 48 hours with a PMI error range from –3 hours to +6 hours. The ADH (base 50 °F) is 4200 °F

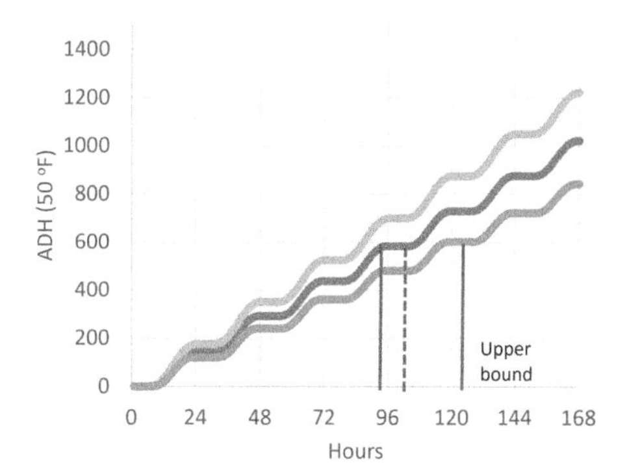

Figure 10.4 Decreasing accuracy of Accumulated Degree Hour (base 50 °F) over time. ADH estimates based on ASOS/AWOS measurement error with increasing time interval from day of body discovery assuming an average daily temperature of 55 °F. The average ADH (orange circle) and upper and lower bounds of the error in ADH (yellow and green circles respectively) are indicated. The corresponding day when 600 ADH is reached is indicated by the blue dashed line. The range in number of days in which 600 ADH might be attained is indicate by the blue solid lines. It is assumed here that there is no modeling error.

hour (corresponding to 175 °F day) after one week with a PMI error range of−9 hours and +14 hours.

But if the minimum daily temperature at an ASOS/AWOS station is less than the base temperature, the undulation over time of the accumulated ADH is accentuated and the error bounds are much broader than when the daily variation in temperatures are all above the base temperature. For example, for days with an average air temperature of 55 °F, an ADH after 48 hours is 291 °F hour (12 °F day) and has a PMI error range of between −7 hours and +15 hours. After one week the ADH is only 1020 °F hour (425 °F day) with a PMI error bound of −28 hours and +39 hours.

The minimum PMI will occur at the lower bound (estimate - error) of the estimated ADD that gives the minimum heat for insect development[8] or corpse decomposition stage.[4, 5] As the observed development stage of a larvae advances or the decomposition is more complete, the error of the estimated ADD resulting from the microclimate of the corpse increases and the PMI estimation error bounds increases.

Other sources of error in PMI determinations

As we have stated, we need to know the microclimate of the corpse over the period of the person's last sighting to the corpse discovery to estimate ADD or average temperature and RH for that microclimate so that a PMI with an error bound can be estimated. A partial list of situations where the PMI and error bound will require the development of different scenarios representing different microclimates for different windows of time include:

- the body being moved after cooling [for example 20% of the corpses in Korean homicides were moved after 2 days[9] and 1/3 of the corpses were moved from the crime scene in North America[10]],
- the body being found close enough to a water body that it may be in or out of water with change in water level or flow rate [for example 25% of corpses in rural Finland[1] and 1/3 of mutilated corpses in Japan[11] were found in water],
- the body being found in a **conditioned** building with unknown thermostat settings [for example 1/3 of the corpses in North America[10]].

References

1. Flores, V., Kim, H., Sielawa, M., Malinowska, P., Ramanauskas, B., Becker, D., LeRoux, H., Häkkänen, H., Hurme, K. and Liukkonen, M., 2007. *Journal of Investigative Psychology and Offender Profiling, 4*(3), pp.181–197.
2. De Matteis, M., Giorgetti, A., Viel, G., Giraudo, C., Terranova, C., Lupi, A., Fais, P., Puggioni, A., Cecchetto, G. and Montisci, M., 2021. *International Journal of Legal Medicine, 135*, pp.193–205.
3. Vass, A.A., 2011. *Forensic Science International, 204*(1–3), pp.34–40.
4. Megyesi, M.S., Nawrocki, S.P. and Haskell, N.H., 2005. *Journal of Forensic Sciences, 50*(3), p.JFS2004017.
5. Moffatt, C., Simmons, T. and Lynch-Aird, J., 2016. *Journal of Forensic Sciences, 61*, pp.S201–S207.
6. Komar, D.A., 1998. *Journal of Forensic Sciences, 43*(1), pp.57–61.
7. Brooks, C.E.P. and Carruthers, N., 1953. *Handbook of statistical methods in meteorology*. Air Ministry, Meteorological Office, M.O. 538. Her Majesty's Stationary Office, London. 412p.
8. Matuszewski, S., 2017. *International Journal of Legal Medicine*, 131(3), pp.877–884.
9. Sea, J. and Beauregard, E., 2018. *International Journal of Offender Therapy and Comparative Criminology, 62*(7), pp.1947–1966.
10. Chapman, B., Raymer, C. and Keatley, D.A., 2022. *Homicide Studies, 26*(2), pp.199–215.
11. Zaitsu, W., 2022. *Journal of Forensic Sciences, 67*(6), pp.2367–2375.

Glossary

Absorptivity The fraction of **radiant energy** reaching a surface that is absorbed by a surface and adding **sensible heat** to the material.

Accuracy A statistical measure of how close a measurement of a condition is to the actual condition. For instance, the measured **temperature** by some instrument will be within so many degrees to the actual temperature 95% of the time it is measured.

Accumulated Degree Days A measure of accumulated heat with units of °F day or °C day. For example, if the average daily temperature on day 1 was 45 °F and that on day 2 was 60 °F, the ADD for day 1 and 2 would be 45 °F day and 105 °F day respectively.

Accumulated Degree Hours A measure of accumulated heat with units of °F hour or °C hour. For example, if the average hourly temperature at 9 AM was 50 °F and that at 10AM was 55 °F, the ADH for 9AM and 10AM would be 50 °F hour and 105 °F hour respectively.

Adiabatic The cooling of air as it rises and expands without any added sources or sinks of **sensible heat**, or the warming of air as it sinks and compresses without any added sources or sinks of **sensible heat**.

Advection The movement of heat or **water vapor** by horizontal water and air movement

above ground level Height above the ground at the specific place of measurement.

above average sea level Average sea level across the entire world.

Automated Surface Observation System A network of meteorological/climatological measurement stations operated by National Oceanic and Atmospheric Administration. Typically located at major airports in the United States.

Atmospheric pressure The cumulative effect of the Earth's gravitational pull on air molecules above a given place. The atmospheric pressure decreases with increasing altitude at a given location.

Automated Weather Observation System A network of meteorological/climatological measurement stations operated by United States Federal Aviation Administration. Located at regional and local airports in the United States.

Base temperature The **temperature** below which insect larval development ceases.

Bias A measure of the average difference between a measurement of a condition and the actual condition. For instance, how different is the measurement of **temperature** by some instrument to the actual temperature.

Bouyancy The tendency for small volumes of air or water to rise because it is less dense than the air or water around it.

Bouyant Tending to rise in height in the air or water.

British Thermal Unit A measure of heat, represents amount of heat required to raise the **temperature** of one pound of water by one degree Fahrenheit.

Climate The characteristic **temperature**, wind and **humidity** typical for a location due to the long-term variation in weather conditions. Typically representing 30-year averages.

Cloud cover The fraction of sky covered with clouds.

Cloud height The altitude of the bottom of cloud cover.

Convective exchange coefficient A number that represents how readily heat is moved by the air or water. Values described with units of BTU per hour per unit area per degree temperature difference (**BTU** hr^{-1} ft^{-2} **°F**$^{-1}$)

Coefficient of determination A number that describes the fraction of variation in predicted values explained by a model given known values.

Cold climate Climates in which the coldest month average air temperature is below 27 °F the warmest month average air temperature is above 50 °F.

Condensation The conversion of gaseous **water vapor** to liquid water.

Conditioned A room or building in which air temperature and humidity are modified by a space heater, furnace and/or air conditioner.

Conduction The movement of heat through solids as a result of a **temperature gradient**.

Conductivity The ability to conduct heat. Units of the amount of heat transferred in an inch of material in an hour across some area per degree temperature difference. Values described with units of **BTU** in hr^{-1} ft^{-2} **°F**$^{-1}$.

Convection The movement of heat with movement of air or water.

Correlation coefficient A number that shows how two variables are linearly related to each other.

Climate Reference Network An automated climate measurement network operated by National Oceanic and Atmospheric Administration.

Data Measurements or statistics of measurements that can be used to describe or understand situations or conditions over time or space.

Database An electronic collection of **records** of measured values or images.

Data logger A device to **record** measurements of a sensor at a regular frequency.

Dew point temperature The **temperature** of the air when water will start to condense out of the air as you cool the air.

Earth radiation The radiant heat emitted by everything on the earth's surface that exists at **temperatures** of the order of 50 **°F**.

European Climate Assessment and Dataset The European database for the collection, qualification, and retention of climate data.
Effective temperature The **temperature** of the air corresponding to the radiant energy emitted by the entire depth of the atmosphere including clouds to a surface.
Emitted/emission To send something out or have something come out from a surface such as light.
Energy balance The mathematical representation of pathways for energy to come into a surface and those going out from the surface.
Error A measure of the difference between a measured value versus the actual value.
Error bound The range in values that the estimated value could be given the possible **error**s in making the estimate. The bound ranges from the estimate plus the error of the estimate to that minus the **error** of estimate.
Evaporation The conversion of liquid water to **water vapor**.
Finished A room in which the structural materials of the walls, floor and ceiling are covered.
Fog Condensed droplets of water in an air layer in contact with the ground.
Forced convection The transfer of **sensible heat** by the wind.
Frequency distribution A distribution showing the number of occurrences of measured or modeled values within fixed ranges of values.
Front A region in which there is a steep **gradient** in air **temperature** and/or **humidity** over the landscape.
Gradient A change in a property over a distance such as **temperature** over a thickness of a wall.
Heat capacity How much heat an object of unit volume increases with increased **temperature**. Values described with units of **BTU** ft^{-3} **°F**$^{-1}$.
Heat content The molecular motion or heat contained in an object.
Humidity The amount of **water vapor** in the air.
Inflitration The natural horizontal flow of air through gaps in solid materials such as forest tree trunks or building doors and windows due to winds.
Insulation Material that reduces the **conduction** of heat.
Lapse rate The rate of change in air **temperature** with altitude or elevation. Values described with units of **°F** per 100 or 1000 feet.
Latent heat Heat that is stored in the structure of water molecules. This heat cannot be measured directly but is calculated based on how much water is in what form (gas, liquid, solid).
Latitude The angle between a line straight up from the Earth's surface at the location of interest that goes through the center of the earth, and a line from the center of the earth that goes straight up from the earth's surface at the Equator: Values range from −90° at the south pole to +90° at the north pole, with 0° at the Equator.

Litter Un-decomposed dead leaves, stalks, bark, cones, and woody branches that accumulate on the forest floor, under crops or on grassland ground surfaces.

Local climate The climate near the ground air that is influenced by small changes in the surface features such as hills, valleys, urban areas, coastal regions. Influences are in time scales of hours to a day within hundreds of feet to tens of miles.

Meteorological Assimilation Data Ingest System A United States database for the collection, qualification, and retention of climate data.

Mean Absolute Error A measure of the tendency of differences between a measured and predicted value to be zero.

Maritime climate The general climate of land masses near the ocean and influenced by nearby ocean **temperatures**.

Mass A large group of organisms with no definite shape.

Mean Bias Error a measure of the tendency for differences between a measured and predicted value to be zero.

Metadata Information about the characteristics of measurement data such as frequency of measurement, location of measurement, how it is measured.

Microclimate The climate associated with variations in the physical and biological environment over time periods of minutes to hours over distances of feet to several 100's of feet and vertically within four times the height of the surface features (trees, buildings, etc). Examples of microclimates include the environmental conditions of roadways, small lakes, cropped and forested land, parking lots, or within buildings.

Maximum Minimum Temperature System Measurement sensor system used at the 'Summary of the day' Cooperative climate stations.

Natural convection The vertical transfer of **sensible heat** due to **bouyancy**.

National Institutes of Standards and Technology an agency of the United States Department of Commerce responsible for measurement standards.

National Oceanic and Atmospheric Administration An agency of the United States Department of Commerce.

National Climate Data Center A center that once collected, stored and disseminated climate data- The climate data is now stored by the **NCEI**.

National Centers for Environmental Information A center that collects, stores, and disseminates climate data. A Center of the of the United States National Oceanic and Atmospheric Administration.

Net radiation The solar radiation and **thermal radiation** absorbed at a surface minus the solar and thermal radiation reflected or emitted from a surface.

Normal distribution A symmetrical bell-shaped distribution of the frequency of occurrence of a collection of numbers sorted by magnitude where the most frequent value is the **mean** value and the values greater and less than the mean value can be described by the **standard deviation**.

National Weather Service An agency of the United States National Oceanic and Atmospheric Administration.

Open terrain Land that is relatively free of trees and shrubs. This includes deserts, large construction sites, sparse grasslands, barren plains, and rocky or grassy hills and mountains.

Post-Mortem Interval The time interval between a person's death and their corpse discovery.

Precipitation Any form of water (liquid, ice or snow) that comes out of the sky and reaches the ground.

Radiant energy Energy transferred without **mass**, such as light.

Radiation Energy that is **emitted** by all objects based on their temperature. Some radiation we see with our eyes and some we can feel with our body such as when we are near a fire or stove.

Record The originally-reported measurement or revised measurement that is provided in a **database**.

Reflectivity The fraction of **radiant energy** that is reflected from a surface.

Relative humidity or RH The ratio of how much water is in the air to how much water the air can hold at its **temperature**.

Resolution The smallest change in the environment that a specific sensor can detect.

Root Mean Squared Error A measure of the variability in individual time-period differences between a measured and predicted value.

Roadway Weather Information Systems Climate stations that measure road and weather conditions along highways and are maintained by United States state highway agencies.

Sensible heat Heat that can be measured by **temperature**.

Sensor A device to measure some property of the environment.

Soil pore The space between individual soil particles. Typically filled with water or air.

Solar radiation The **radiant energy** emitted from the sun.

Specific humidity A measure of water vapor as the weight of the water per weight of air.

Standard deviation A description of how variable a measure is assuming a random probability distribution. 68.3% of all values should be +/- the average value.

Sky radiation The **radiant energy emitted** from the sky and clouds in the sky due to the temperature of the air and the clouds.

Sky view The portion of the sky that can be seen from a surface looking upwards in all directions.

Still air Air that is not moving.

Sublimation The change from ice to **water vapor**. Associated with an increase in **latent heat**.

Synoptic weather A snapshot of large-scale **weather** conditions across an expanse of space for a specific instant in time showing regions of high and low **atmospheric pressure**, **fronts**, and regions of **precipitation**, and **cloud cover**.

Taphonomical The knowledge of the processes causing changes in the human body after death.

Temperature A measure of **sensible heat** content of a material.

Temperate climate A moderate **temperature** middle latitude (between tropical and polar latitudes) climate where the temperature variation over the year is more than the temperature variation in a day. There are distinct summer and winter seasonal temperatures and **precipitation**.

Thermal radiation **Radiant energy** emitted by everything on the earth's surface and in the sky that exists at **temperatures** between approximately –50 °F to 150 °F. This includes clouds.

Time constant The amount of time that it takes for a sensor to detect 64% of a change in the environment that the sensor is measuring.

Understory The layer of vegetation between the bottom of the forest tree canopy reaching the top of the forest and the ground. Commonly contains small trees and shrubs.

Unconditioned A room in which air temperature and humidity are not modified by a furnace, space heater, and/or air conditioner controlled by a thermostat.

Unfinished A room in which the walls, floor and ceiling show the structural materials.

Ventilation The mechanical flow of air through buildings and between buildings and the surrounding air.

Water vapor Water in the gaseous state.

Weather The condition of the atmosphere (such as air **temperature**, pressure, **humidity**, **precipitation**, and winds) at any given specific time and location.

Appendices: Case studies

Case 1: Corpse discovered in scrubland in winter

The partially-decomposed body of a young female was found in New Mexico under a bush in scrubland near an earthen dam on a river at 3050 ft. asl on 30 January 1982 (Figure A1.1). Testimony claimed that she was struck on the head, walked out on an earthen dam then fell down a hill into grass and weeds and left to die from exposure. The girl was last undisputedly sighted on 21 December 1981. The conviction was based on an assumed murder on 1 January 1982 after a party.

Approach

Estimate the ADD for the time period of last sighting of the female to 30 January 1982 when the body was discovered using time of observation and elevation corrections and spatial interpolation (Chapter 7) of multiple climate stations varying in distance and elevation from the body.

Figure A1.1 Vegetation and terrain surrounding the body discovery location. The location of the corpse discovery (red circle) is indicated.

Imagery from USDA National Agricultural Image Program.

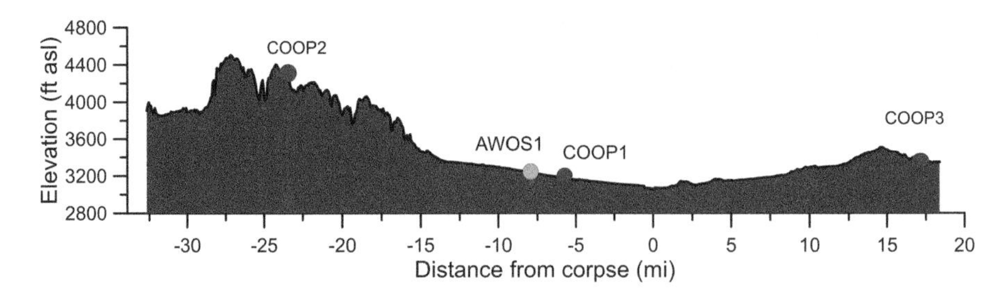

Figure A1.2 Profile of terrain surrounding the body discovery location. Nearby 'Summary of the Day' climate stations (blue dots), an AWOS airport station (green dot) and the location of the body (zero distance from the corpse) are indicated.

Sources of data

- Records of 'Summary of the Day' measurements made at 0800 LT at COOP1 (3120 ft asl elevation and 5.5 miles from the location of the body), COOP2 (4422 ft asl elevation and 21.8 miles from the location of the body) and COOP3 (3524 ft asl elevation and 17.4 miles from the location of the body) (Figure A1.2).
- Hourly weather record at AWOS station AWOS1 recorded typically between 0600 and 1800 LT (3232 ft asl elevation and 7.8 miles from the location of the body) (Figure A1.2).

The COOP1 station was located to the south of the city-center in a low-density industrial region. The COOP2 station was located on a plateau in open land near buildings. The COOP3 station was located on relatively flat land of scrub desert. Of the available daily temperature records, the COOP1 station was closest in elevation to that of the body on discovery (5.5 mi). Of the three

COOP stations, the COOP1 station elevation was the closest (70 ft) to the location of the body discovery. Objectively, the best estimator of the conditions at the body would be measurements made at the COOP1 station.

Modeling

Unfortunately, the maximum and minimum temperature observations at COOP1 were only partially present or entirely missing on 13 days over the two-month period in question while COOP2 and COOP3 stations had a complete set of observations over the period of record (Figure A1.3). Furthermore, one day's temperature maximum, minimum and time of observation record was inconsistent and invalidated at COOP1. There was also a day of missing AWOS1 observations. Since daily temperature observations were made at COOP1, COOP3 and COOP2 climatological stations at 8am, the maximum temperature records for all locations were shifted back one day. This shift resulted in five additional days of incomplete maximum and minimum daily temperature observations at COOP1.

The AWOS1 station recorded hourly temperatures during the hours of operation (typically 6 am to 6 pm LT although somewhat variable during the period of time.) Since a typical clear sky has the minimum air temperature just before sunrise and the maximum temperature in the afternoon, the minimum and maximum of the recorded temperatures during a given day were used as the minimum and maximum daily temperatures of that day.

Estimated missing COOP1 station average daily temperature from the average daily air temperatures reported or estimated at AWOS1, COOP2 and/or COOP3. The elevation-corrected daily average temperatures of AWOS1 and COOP3 stations were closer to that of COOP1 than the COOP2 station. The average difference in the elevation-corrected daily temperatures at COOP1 and the average daily temperature based on AWOS1 and COOP3 stations over the two months or record was only 0.3 °F +/− 3.4 °F, indicating good agreement between the daily average temperature records these three locations (Figure A1.3).

Conclusion and potential impact of errors

Since the body was discovered at an elevation of approximately 3050 ft asl, the average daily temperatures at the body were estimated by elevation-corrected spatially-interpolated average temperatures using daily average temperature records at AWOS1, COOP1, and COOP3. The ADD for the period of 21 December 1981 (day of last undisputed sighting of the female) to 30 January 1982 was 217 °F day with an error bound of 191 °F day to 244 °F day. The ADD for the period of 1 to 30 January 1982 was 149 °F day with an error bound of 130 °F day to 169 °F day.

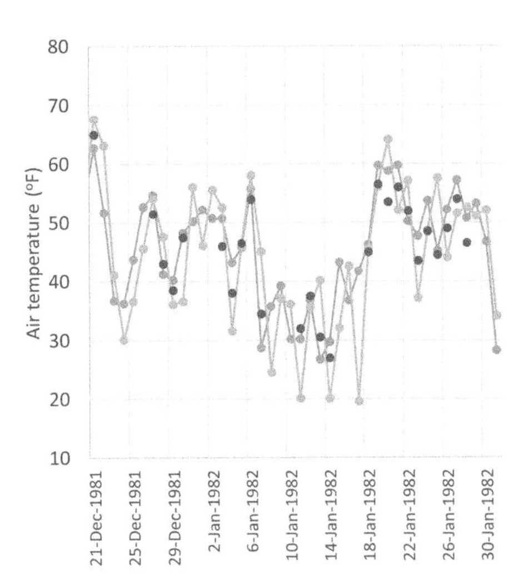

Figure A1.3 Comparative average daily air temperatures for three climate stations near the body discovery location. Elevation corrected daily average air temperatures at AWOS1 (yellow circle and line), COOP1 (blue circle and line) and COOP3 (grey circle and line) are indicated.

Data source: NOAA National Center for Environmental Information https://www.ncei.noaa.gov/access

Case 2: Corpse discovered in a forest in winter

A female college student was murdered and her body dumped in brush in a heavily wooded area in Texas at an elevation of 214 ft asl and 215 ft from a 22,000 acre lake (Figure A2.1). The face and neck had undergone extensive decomposition while other exposed portions exhibited much less decomposition. The investigation was conducted by the local County Sherriff's Office. She was last seen on 7 December 1998. Hunters found the female's body the morning of 2 January 1999. The defendant was arrested for evading arrest or detention on 11 December 1998.

Approach

Estimate the ADD for the corpse on the forest floor using a linearized energy budget in combination with hourly records of a nearby weather station. Estimating the accumulated heat at the deceased body surface in the forest required: A) the heat accumulation on an hourly basis, B) adjustment for the difference in elevation between the measurement location and the location of discovery, and C) adjustment of hourly air temperatures for the effects of the forest canopy over the body.

Figure A2.1 Location of female corpse. The corpse location is indicated by the red circle.

Source: US Geological Survey https://www.usgs.gov/programs/national-geospatial-program/topobuilder

Sources of data

- Records of 'Summary of the Day' measurements at nearby COOP stations COOP1 (245 ft. asl and 16.7 mi from the location of discovery) and COOP2 (494 ft. asl and 14 mi from the location of discovery.)
- Hourly weather records at ASOS station ASOS1 (245 ft. asl and 17 mi from the location of discovery.)

The distance between the location of discovery and measurement station was approximately the same for all three stations although the COOP2 station was closest. None of the measurement stations were near a water body. COOP1 and ASOS1 station were both at the same elevation and nearest the elevation (31 ft) of the location of discovery.

Given the variable influence of the forest and lake on the forest floor climate during the course of the day, the microclimate of the forest floor and resulting ADD were calculated from the hourly temperature records at ASOS1 rather than the daily maximum and minimum temperatures from the COOP stations. The elevation difference between the location where the body was found and the ASOS1 station corresponded with a temperature correction of +0.2 °F.

Modeling

Based on the photographs and aerial imagery, the forest where the body was found was a relatively dense forest of mixed conifer/deciduous trees with heavy understory vegetation. The canopy density was relatively high with the deciduous species senesced due to both tree boles and the mix of coniferous species. There was an approximately 30% view of the sky which limited the radiational heating and cooling of the air at the forest floor. In addition, the body was found directly under a tree, further limiting the view to the sky. Although the body was found near a lake body, we assumed no lake influence on the average temperature given the winter season.

A microclimate model was developed from the hourly ASOS1 station records to describe the hourly forest floor temperature. The winds, cloud cover, and air temperature over the forest were assumed to be the same as the ASOS1 measurements. The modeled forest floor minimum daily air temperatures were often higher and the maximum temperatures were occasionally lower than ASOS1 (Figure A2.2).

Conclusion and potential impact of errors

The ADD (base 43°F) of air over the forest floor over the 8 December 1998 and 2 January 1999 period of record was 208 °F d. The propagation of the ASOS1 temperature measurement error in the ADD adjusted for forest influence

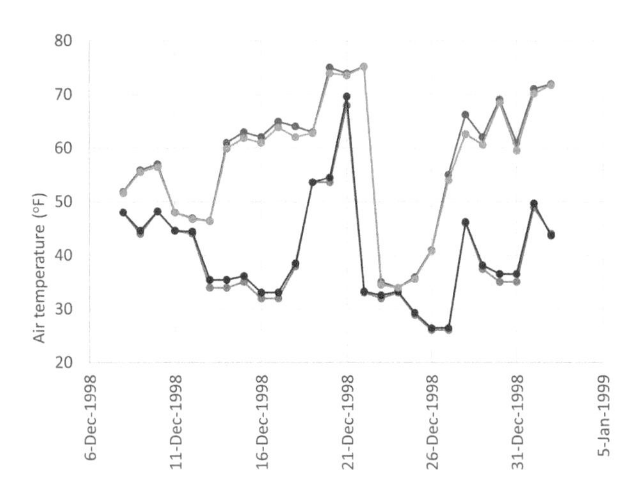

Figure A2.2 Time history of temperature record and modeled temperature at corpse. The measured daily maximum (blue circle and line) and minimum (orange circle and line) temperature at ASOS1 and the modeled daily maximum (green circl and line) and minimum (purple) temperature at the corpse location.

Data source: NOAA National Center for Environmental Information https://www.ncei.noaa.gov/access

resulted in a range of 176 °F d to 226 °F d, less than that based on the COOP2 station alone. If the forest influence was ignored, the ADD based on the COOP2 station was 20 °F d higher than when the forest influence was considered.

The ADD for the sub-period of 11 December 1998 and 2 January 1999 was 177 °F d adjusted for forest influence with a propagated measurement error bound of 156 °F d to 198 °F d. Entomological evidence led the forensic entomologist to conclude that the body was colonized by *C. cadavarina* between 5 December and 9 December 1998. Based on the modeled ADD, the colonization could have occurred as late as 12 December 1998. The ADD error bounds of the two time periods precluded the ability to assess whether the female was murdered on the last day she was seen or the day that law enforcement arrested the defendant.

Case 3: Multiple corpses discovered in a forest cabin in winter

A male and his live-in girlfriend were murdered by shotgun in a cabin at 7500 ft asl within a National Forest in New Mexico in winter or early spring. The locked cabin door confined their two young sons who died at a later time of starvation and dehydration. The bodies of all four were discovered on 14 April 1996. The State Police and detectives from the nearby Police Department investigated the four deaths. The last date that the man was seen was 10 December 1995.

Approach

Estimate the ADD within the cabin for the two shotgun homicides and the starvation of the two children by modeling the energy budget of the cabin utilizing nearby climate records. The model included: A) cabin infiltration based on monthly average wind conditions, B) outside air temperatures using daily average air temperatures adjusted for the difference in elevation between the measurement location and the location of discovery, and C) modeled daily equilibrium temperature differences between the within-cabin microclimate and the above forest canopy climate using daily average air temperatures.

Sources of data

- Records of 'Summary of the Day' measurements from observations at 4 pm LT using liquid-in-glass thermometers at the COOP1 Station (6520 ft elevation; approx. 14 miles from the cabin on the same side of the mountain range as the cabin).

- Hourly weather records at the ASOS station ASOS1 (5310 ft elevation; approx. 23 miles from the cabin on the opposite side of the mountain range as the cabin).

The COOP1 station (Figure A3.1) was situated in rolling hills with numerous stands of cedar, Juniper and Pinion. Surrounding terrain can influence the air temperature measurements. The minimum daily temperature record over the period of interest had one recording error (recorded minimum temperature was greater than the temperature observed at the time of recording) while the maximum temperature record had five errors (prior day temperature at observation was greater than the maximum temperature). The errors were assumed to be due to the inadequate resetting of the maximum and minimum thermometers since the maximum and minimum thermometers needed to be shaken down by the observer to reset after each observation. Therefore, the correction for the daily maximum temperature error was assumed to be that of the prior day current temperature at recording and the correction for the daily minimum temperature was assumed to be the current temperature at recording.

Adjustments to climate record

The temperature outside the cabin was estimated using the altitude-corrected temperature record from the MNM COOP weather station. The cabin had a light-colored roof and adobe-colored outer walls. Based on investigators, the trees were dense to the east next to the cabin. Although 50% of the sky view was obscured by trees, solar radiation heating was not estimated because the cabin was reported by investigators to be well-insulated.

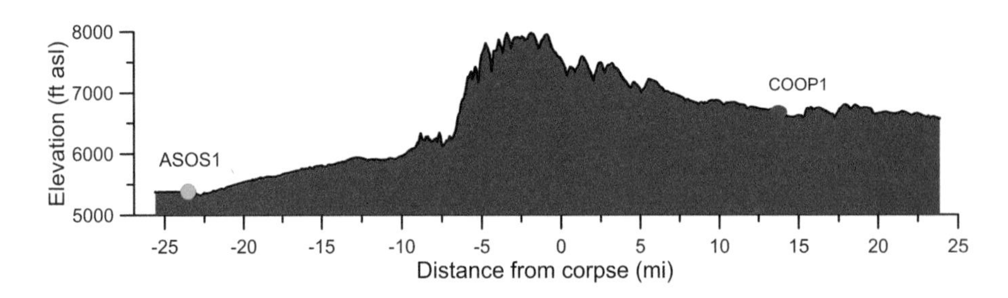

Figure A3.1 Terrain profile for climate information sources and cabin location. Cabin (red circle) and nearby stations with temperature records: COOP1 station (blue circle) and an ASOS station (Green circle) are indicated.

Modeling

The total floor area of the cabin was 550 ft^2 with the children's bedroom was 154 ft^2. There were three windows of which two were open: The window in the children's room and the partially-open window in the parent's bedroom. The assumed height of the ceilings was 7 ft. Total surface area of walls and ceiling were 651 ft^2 and 550 ft^2 respectively.

Heat gain in the cabin included two 1500W electric space heaters running in the cabin when the bodies were discovered. The space heaters running in the cabin room each produced 1.3 MCal per hour. Assuming no ventilation in the cabin, this heater was estimated to heat the entire cabin interior at a rate of 1.3 min per °F.

Heat loss through the walls and ceiling were estimated assuming well-insulated conditions. The heat loss through the walls and ceiling for the entire cabin was estimated at 56 calories per degree F difference between the inside and outside temperatures per hour.

There were three windows open (as discovered by investigators) in the cabin. Infiltration through the cabin windows was estimated based on the wind speeds at ASOS1 and canopy cover. Although measured wind speeds at ASOS1 (5310 ft asl) were likely lower than those above the canopy at the cabin (7500 ft asl), the average monthly wind speed at ASOS1 was used as a rough proxy for the above-canopy winds with the expectations that the infiltration was probably underestimated. Winds at the height of the cabin windows was approximated as 20% of above forest wind speeds. Given the average monthly wind speeds at ASOS1 of 9 mph, the average wind speed at the cabin influencing the infiltration of air into the cabin was estimated at 2 mph. It was estimated that there were between three and eight exchanges of air between the interior and exterior of the cabin per hour.

The equilibrium temperature of the air in the house (whole house or children's room) was estimated (assuming good mixing of air in the house) based on four components to the heat budget for the house (heat transfer through the walls and ceiling, heat transfer through the floor, and transfer through the open windows, and heat transfer from the two 1500W electric heaters. The equilibrium temperature was determined to be 18 °F above the outside air temperatures. A change in the difference in air temperature within the cabin and outside the cabin would take only 24 minutes to come to an equilibrium. If the children's room door was closed the children's room equilibrium would take only 4 minutes as a result of the open window. Excluding the heat exchanges across the walls, ceiling, and floor only changed the equilibrium temperature by 0.4 °F. The child found on the top bed of a bunk bed would be expected to be at equilibrium with the air in the cabin.

Although the adult male, female and one child's bodies were on the floor, and floor temperatures in the house can be expected to lag that of the air by one to two days (assuming light construction and temperature equilibration at 95% of the ambient air temperature). Given the period of time under consideration, this lag was not considered in the final analysis.

Conclusion and potential impact of errors

The adjustments of daily average temperatures from COOP1 included a −5.5 °F temperature adjustment for elevation for air outside the house and a +17.6 °F temperature adjustment of air in the cabin relative to that outside the cabin. The ADD (base 43°F) for the period 10 December 1995 to 14 April 1996 was 1238 °F d with bounds of between 107 °F d lower to 155 °F higher. Since no estimate of radiation heating of the well-insulated cabin from the sun was included in the analysis and the winds at the cabin were likely underestimated, the estimated within-cabin temperatures were likely a slight underestimate. The decomposition of the bodies and the entomological evidence in combination with the ADD information brought the medical examiner to conclude that the two adults died on or around 12 December 1995 and the two children died sometime in early January 1996.

Case 4: Corpse discovered in an unconditioned garage in the spring

A male corpse was found in Kansas in an unheated outbuilding under a tarp held down by cinder blocks. The male corpse was found on 6 May 2013 by authorities after being alerted by friends of the deceased. The man was murdered by shotgun. The man was last seen 18 April 2013.

Approach

Estimate the ADD for the male corpse using a simplified energy budget of the outbuilding (968 ft asl) based on post-discovery measurements and nearby hourly climate records. Modeled temperatures within the outbuilding assumed the main door to the building was open and were dependent on solar radiation, wind, and outside air temperatures. The model was calibrated to the specific outbuilding using the post-discovery measurements.

Sources of data

- Hourly ASOS measurement record at ASOS1 (1075 ft. asl, 39 mi from outbuilding) and ASOS2 (1060 ft asl., 30.5 mi from outbuilding).
- Records of 'Summary of the Day' measurements at station COOP1 (1063 ft. asl; 9.4 mi from outbuilding) and COOP2 (900 ft. asl., 6 mi from outbuilding).

- Hourly measurement record at a state-operated MesoNet station MESO1 (969 ft asl, 9.5 mi from outbuilding).
- Hourly record of 10-day post-discovery experiment measuring the hourly air temperatures inside and outside the outbuilding conducted by the local County Sherriff Criminalistics Laboratory
- Solar Position chart location

Assessment and adjustments of measurement errors

- Air temperature sensor at MESO1 did not have a calibration protocol but did report to have infrequent undocumented inter-comparisons made of sensors during site visits. The sensor was located near a small airport and had a fetch of greater than 100 ft. These records were used for prediction of the air temperature in the outbuilding/outbuilding. The station is located 30 ft from buildings and 160 ft from a line of trees to the north.
- The humidity sensor used at MESO1 had a manufacturer-supplied error of +/− 3%.
- Air temperatures at COOP1 and COOP2 were taken at 7am at both stations. Recorded maximum temperature was shifted back one day. The COOP2 instrument was within 20 ft of a 61 ft building that probably affected the reported temperatures.
- Air temperature sensors used in on-site experiments by the local County Sherriff Criminalistics Laboratory had a manufacturer-reported sensor accuracy of +/ 0.1°F. The sensor calibrations were verified yearly by NIST-traceable liquid-in-glass thermometer. The radiation error associated with the outdoor exposure of these sensor was unknown although effort was made to prevent exposure to direct sunlight. Indoor exposure of the sensor would have a minimal impact on the measured temperature.
- The humidity sensor used in the on-site experiments by the local County Sherriff Criminalistics Laboratory had a manufacturer-reported sensor accuracy of +/− 3%.

Temperature exposure errors between the radiation shield used at COOP1 and COOP2 and the temperature/humidity sensors used at the MESO1 hourly weather station typically result in the daily minimum temperatures 0.1°F lower and daily maximum temperatures 0.5°F higher at COOP1 and COOP2 than MESO1. Therefore, a minimum and maximum temperature difference (COOP2 – MESO1) of +1.2°F and −1.5°F respectively could still represent measurements of the same air.

The need to model the temperatures within the outbuilding required hourly temperature records. Unfortunately, ASOS1 and ASOS2 with hourly measurements were much further away than COOP1 or COOP2. COOP1 was near a lake resulting in air temperatures likely affected by the water body. Since this expected lake influence that would not be present at the unheated outbuilding, the COOP1 records were invalidated for primary use. The nearest station, COOP2, had one day with inconsistent air temperatures – probably a result of the maximum/minimum thermometer not being reset on the prior day. The temperature records for COOP1 were substituted in the COOP2 temperature record since the MESO1 winds were not coming across the lake to the COOP1 station on this day.

The complete daily temperature record at COOP2 was then used to evaluate the uniformity of maximum and minimum temperatures across the area – answering the question: can the more distant MESO1 station describe conditions at COOP2. On average, a comparison of the COOP2 and MESO1 measurements showed excellent agreement with linear correlations between the daily maximum and minimum temperatures were 0.98 and 0.93 respectively (see Chapter 7). The daily temperature range (maximum temperature – minimum temperature) in the hourly MESO1 measurements were comparable to the daily temperature range recorded at COOP2. The difference in MESO1 and COOP2 measurements were within the measurement errors at the two locations.

The terrain surrounding the outbuilding was rolling hills. The difference in elevation of the outbuilding and the MESO1 station was negligible.

Adjustments to climate record

Based on the photographs provided and the comments of the investigators, the 36 ft. by 21 ft outbuilding was non-conditioned with corrugated steel roof and walls oriented east-west, with the main access and pedestrian door opening on the south side. The roof was approximately half-covered with tree branches from a line of trees to the north of the building. There were no leaves on the trees at the time of the discovery. It was assumed that the temperature of the tarp-covered body was in equilibrium with the air inside the building.

The outbuilding inside temperature was modeled accounting for heat exchanges with the roof from the sun, radiant cooling and heating between the building and the air (thermal radiation), heat conducted through the roof and walls, and heat transported through the door by the wind (infiltration). As expected, the air volume of the building was rapidly exchanged with the air outside, indicated by the small difference in humidity between the building and outside during the on-site measurements.

A 10-day period of on-site post-discovery measurements of the air temperatures inside and outside the outbuilding was conducted by the local County Sherriff Criminalistics Laboratory during a similar period of time during the year as the period of interest. These measurements were used to develop the microclimate model. This model was then used to estimate the hourly air temperature in the building during the period of interest using the hourly air temperature, solar radiation, and wind speed record of the MESO1 weather station.

The model predicted the air temperature within the outbuilding during the 10-day on-site period of measurements in 2015 with high accuracy (Figure A4.1). The average error in predicting the average daily temperature outside the building was −0.4 °F. The modeled temperature in the afternoon and evening of the 17th and 20th of March were the only times in which the model did not predict the measured temperatures well (Figure A4.1). The inside air of the outbuilding averaged 3.6°F warmer than the MESO1 weather station.

The estimated temperatures in the outbuilding during the period of interest resulted in an average daily minimum temperature in the building that was 0.8 °F cooler than the outside measured at COOP2 and maximum

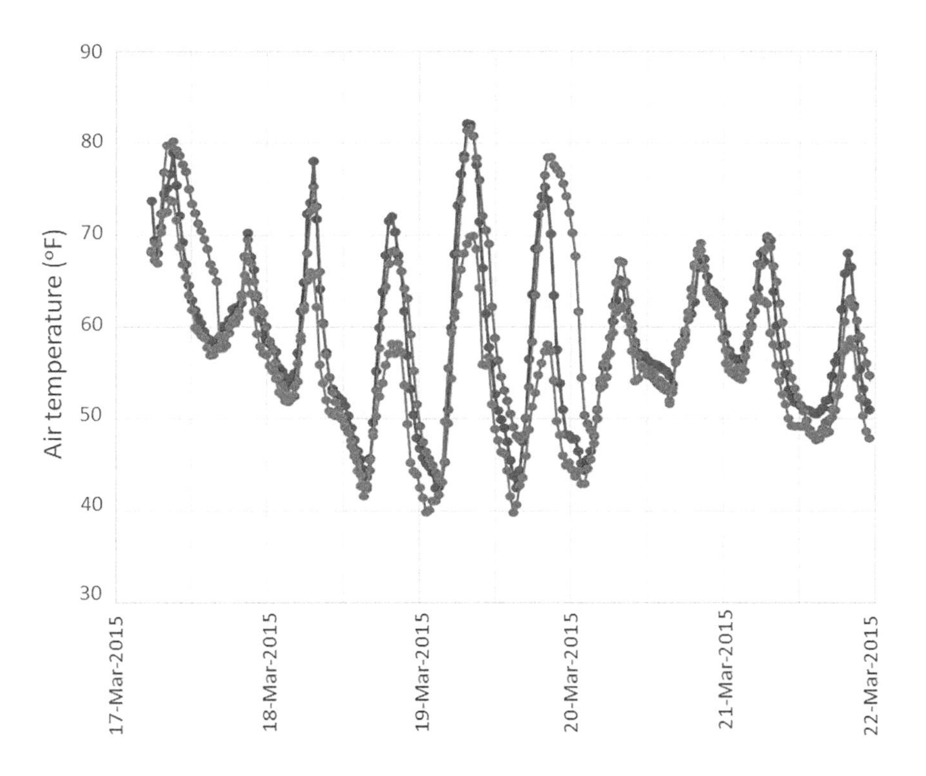

Figure A4.1 Measured and modeled air temperatures during post-discovery experiments. The hourly air temperatures measured at MESO1 (blue circles and line), measured inside the outbuilding (red circle and line) and modeled inside the outbuilding (green circle and line) are indicated.

MESO1 data source: Kansas State University MesoNet, https://mesonet.k-state.edu/weather/historical/. Post-discovery measurements made by local law enforcement.

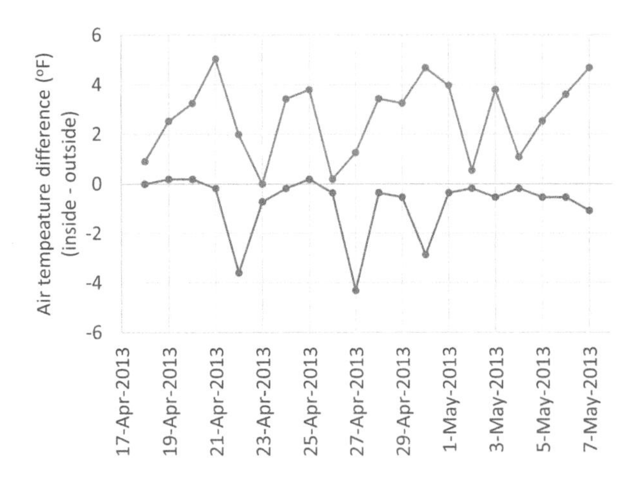

Figure A4.2 Air temperature differences estimated for outside and inside the unheated outbuilding during the period of interest. The daily difference between COOP2 and the modeled inside temperature for daily minimum (orange circle and line) and maximum (blue circle and line) temperatures are indicated.

COOP2 data source: NOAA National Center for Environmental Information. https://www.ncei.noaa.gov/access

temperature that was 2.7 °F higher than outside measured at COOP2. The average daily air temperatures was 3.5 °F warmer in the building than outside – very similar to that measured during the on-site measurements (Figure A4.2).

Conclusion and potential impact of errors

Seven of the 20 days between last sighting and discovery of the body had average daily temperatures in the building below the 43 °F base temperature. The ADD estimate based on the MESO1 climate station was 165 °F d for the period of record. The ADD estimate using the modeled air temperature in the outbuilding was 186 °F d.

Case 5: Multiple corpses discovered in a conditioned manufactured house in the spring

A male and female corpse were found by investigators in a double-wide manufactured house on 6 May 2013 in Kansas. The woman's mother said she last saw her daughter the morning of May 1. The last day that the male was seen was 18 April 2013. They were both found in a SW bedroom by authorities after being alerted by friends of the deceased. On arrival officers found the window in the bedroom closed but the window in the kitchen open. The male was covered with clothing and lying on a heat vent and the female was lying on the floor. A large mass of maggots was found on the male corpse while no maggots were found on the female corpse. Investigators retrospectively believed that the door to the room was open at time of discovery. There was no record of whether floor vents were open or closed at the time of discovery. Furthermore, it was not known whether the male corpse was covered with clothing or lying on the heating vent since his death or whether he was covered just prior to the female's disappearance and death.

Approach

Estimate the ADD within a house for the surface of both male and female corpse under various scenarios to account for a range of unknown conditions during the periods of interest (18 April to 6 May 2013 for the male corpse and 1 May to 6 May 2013 for the female corpse). Scenarios included whether the windows in the bedroom were open or closed and whether the furnace and furnace blower was on or off. The relationship of the measured air temperature at a nearby climate stations to the air temperature in the bedroom under each scenario was assessed using post-discovery measurements. These relationships were then use to estimate the air temperature affecting both corpses under the different scenarios during the periods of interest.

Sources of data

- Records of 'Summary of the Day' measurements at stations COOP1 (900 ft. asl.; 6 miles distant) and COOP2 station (1063 ft. asl; 9.4 mi distant).
- Hourly ASOS measurement records at ASOS1 (1075 ft. asl; 39 miles distant) and ASOS2 (1060 ft asl.; approximately 30.5 mi distant).
- Hourly Mesonet measurement records at MESO1 (969 ft asl; 9.5 mi distant).

The terrain surrounding the house was rolling hills. The COOP1 station was in close proximity to a lake and likely affected by the lake. COOP2 was within 20 ft of a 61 ft tall building that may have elevated the reported air temperatures during the day. One day of recorded air temperatures at COOP2 were in error – probably a result of the maximum/minimum thermometer not being reset on the prior day. The missing day of record was substituted with that at COOP1 since the MESO1 winds were not coming across the lake to the station on this day.

Adjustments to climate record

The difference in elevation of the house (968 ft. asl) and the COOP2 station (900 ft. asl) results in an approximately 0.4 °F correction due to difference in elevation. This 0.4 °F correction to the COOP2 temperature measurements was assumed negligible since the combined effect of measurement error of the temperature measurements (+0.6°F) was greater than the correction.

The 1211 ft^2 north-south oriented house was insulated and the crawlspace was solid-skirted. Several trees were located east of the house with the tree crowns cover about 75% of the house roof. The house was centrally heated. It was assumed that the tree cover and the insulation of the house made outside sources of heat contribute less to the air temperature in the house than the furnace.

The thermostat was located in the kitchen (Figure A4.2). There were at least nine vents for the forced air from the furnace, two of which were located in the SW bedroom (173 ft^2) where both bodies were found (Figure A4.2). Air temperatures in the SW bedroom with the female corpse on the floor depended on whether or not: 1) the room was actively heated/cooled, 2) the room window was open or not (providing infiltration), and 3) the room door was closed or open (providing ventilation with the rest of the house). The male was found under clothing and over a heating vent. The male corpse temperature was assumed to be affected by any active heating or cooling and reach ambient air temperature of the room after some period of time if no active heating were occurring.

The thermostat was set to 75 °F/cool at the time of discovery. But the actual thermostat setting and heating at any other time over the period of interest was unknown. Due to lack of verifiable conditions in the house relative to the furnace operation, the number and location of open vents, windows, and doors and a lack of confidence in the post-mortem body position of the male over the vent prior to discovery, no detailed model of the energy budget of the house was conducted. However, a series of post-discovery experiments were conducted by the local County Sherriff's Criminalistics Lab to get a handle on the variation in temperatures possible in the house. These experiments used three thermometers: one in the shade of the porch outside, one in the room on the bed and one on the vent where the male corpse was found. The errors associated with these sensors were well-documented:

- Reported accuracy of the two handheld units used in the experiments were +/ 0.3°F and +/ 0.4°F; verified yearly by NIST-traceable liquid-in-glass thermometers.
- An inter-comparison of the sensors showed all sensors read the same temperature.
- The radiation error of the sensor on the porch was unknown although effort was made to prevent have the sensor in the shade. No radiation error was likely for the two indoor sensors.

Two extreme scenarios (A and B) to bracket the range of air temperatures that the two corpses could have experienced were developed based on the results of the on-site experiments:

Scenario A: All windows closed, heater blower on/heat on: The thermostat was set to 90°F with outside air temperature of 65 °F and winds 0–7 mph The maximum bedroom air temperature was 3.6°F above the thermostat temperature with the vent temperature equilibrated at 52.8°F above the thermostat temperature. Adding a clothes pile over the vent resulted in an equilibrium temperature at the vent of 11.7°F above the thermostat and the ambient temperature of the room 2.6 °F above the thermostat. This showed that blocking the vent shifted the proportion of air heating the room (as expected given at least eight additional vents of which one additional vent was in the SW bedroom) and that the insulation of the clothing over the male body resulted in a temperature under the clothing 9.1°F above the air temperature of the room. As a result, we estimated that the average daily temperature of the air in the bedroom would be +3.6°F above thermostat temperature for the female corpse exposed to the ambient air and +11.1°F above the ambient air (14.7°F above thermostat setting) for the upper surface of the male corpse when buried under clothing over the vent.

These warmer conditions could account partially for the large maggot mass on the male corpse. Although the presence of clothing on top of a corpse may result in heat being retained at the corpse from the metabolic activity of the maggot mass, this heat contribution was unknown and not estimated. Consequently, the average daily temperature for the male body covered in clothing and heated by the HVAC vent under the body was assumed to be equivalent to that without the metabolic heat from the maggot mass. The minimum and maximum temperatures for the male body covered with clothing may be assumed to be equivalent to that of the thermostat.

Scenario B: All windows open, heater blower on/heat off: All open windows provided a cross-flow ventilation based on wind speed and air pressure. Average wind speeds were moderate at 7.5 mph from the north (based on MESO1) providing light ventilation of the house due to the tree line to the north. The outside air temperature ranged from 44 °F to 65 °F, spanning the minimum and maximum temperatures at MESO1 for the day of on-site measurements. The air temperature in the bedroom averaged 1.5 °F higher than the outside temperatures, ranging from −3°F during the day to +3°F during the night during any given hour. This higher temperature was likely due to the low radiative heat loss in the room compared to outside. In addition, the HVAC blower itself heated the air 0.5 °F. It is reasonable to assume that the equilibrium temperature between the house and the outside would be slightly elevated due other appliances such as refrigerators being heat sources. Furnishings in the house store heat which was the likely cause for the time-shifted temperature response of the house to the outside environment: the minimum and maximum temperatures in the house were not at the same hour as that outside.

These temperature differences were applied to the maximum and minimum temperatures reported at COOP2. The minimum temperature in the bedroom during the experiment was 0.3°F higher than the outside temperatures while the maximum temperature was 2.2°F lower than the outside temperatures. These temperature differences, representing only one-half day of measurements with winds from the most wind-speed-limiting direction, were likely minimal values due to cross-ventilation in the house since all windows were open.

Conclusion and potential impact of errors

A range of estimated temperatures within the house were defined given the lack of knowledge about the specific positions of windows and the use of the furnace:

- If the furnace thermostat was set at 75 °F/cool and the windows were closed, a body over a vent and under clothing in the SW bedroom (similar to that found during discovery) will be approximately 11.1°F above the bedroom air and 14.7°F above the thermostat setting.
- If the furnace thermostat was set at 75 °F/cool and the windows were closed, a body found in the bedroom exposed to the bedroom air will be approximately 3.6°F above the thermostat setting.
- If the furnace thermostat was off and house windows open, a body found in the bedroom exposed to the ambient air or under clothing will have an average minimum temperature approximately 0.3°F higher than the outside temperatures and an average maximum temperature 2.2°F lower than the outside temperature.

For the period of 1 May through 7 May the ADD for the female corpse was 27 °F day if the thermostat was not on and 183 °F day if the thermostat was set to 70 °F / heat. Since higher temperatures in the room would promote maggot development on the female corpse, and maggots were not found on the body, it appears that the thermostat setting of 75 °F /cool could have been the setting during this period.

For the period of 18 April through 7 May the ADD for the male corpse upper surface either exposed to the ambient air of the room or under the clothing was estimated at 131 °F day if the thermostat was not on. If the thermostat were set at 75 °F / heat the ADD for an exposed male body was estimated at 617 °F day. If the male body was under the clothing and over the vent with the thermostat set to 75 °F / heat the ADD for the entire period would be 839 °F day. These ADD estimates were likely underestimates if the male corpse was under the clothing the whole time since the maggot mass heat production is not considered.

Case 6: Multiple corpses discovered in an unheated storage unit during fall

A male and female were murdered by gunshots. The male was dismembered and portions of the male were found in and around a nearby lake on 11 and 12 October 2002. Through monitored phone calls of the convicted murderer, it was determined that a storage unit was rented on 10 October for some undefined use. When the investigators located and entered the storage unit on 17 October, the smell of decay was evident and maggots were evident under the door of the unit. On opening the unit, two large sealed garbage containers contained the bodies of the woman and most of the body of the man. Insect pupae were found inside the container of the male body. Insect larvae were found inside the container of the female body. The last sighting of the couple was 4 October, 2002. The defendant told his ex-wife that he murdered the couple October 6.

Approach

Estimate the ADD for the two corpses in enclosed containers within the storage unit (1650 ft asl) over the period of interest (4 October to 17 October 2002) using nearby climate records to estimate the air temperatures outside the unit and post-discovery measurements to model the storage unit microclimate.

Sources of data

- ASOS hourly weather records at ASOS1 airport at 1491 ft asl elevation and 11.7 mi from the storage unit.
- Daily weather maps

Adjustments to climate record

The difference in elevation between the ASOS1 and the storage unit was 159 ft, the difference would nominally cause the air outside the unit to be 0.9 °F lower. Since the storage unit was within the urban center, a local urban heat island would be expected- resulting in temperatures higher than the nominal temperatures in open fields such as the ASOS1 airport. The magnitude of this urban local climate influence was unknown. Consequently, no elevation correction was made to the airport temperature measurements. Unlike air temperatures, the air moisture content is more constant across the landscape. The only likely moisture sources in the city during the winter were evaporating water from melting snow and central heating exhaust from buildings.

The uninsulated storage unit, with door facing to the NNE and adjoining units on both sides, had an 8 ft ceiling and a volume of 400 ft^3. The major sources of heat that would influence the air temperature in the unit included: 1) sunlight during the day heating the walls and roof with that heat conducted into and out of the unit, 2) radiative heat loss during the night when the sky is clear cooling the roof and walls with heat conducted out of the unit through conduction across the walls and ceiling, 3) radiant heat exchange with the concrete floor removing heat from the air when the concrete is colder than the air and adding heat to the air when the concrete is warmer than the air, convection of heat into or out of the unit when the door was open.

The bodies found in storage containers in the unit would over some period of time be similar to the air temperature as moderated by the concrete floor temperature. The temperature of the storage containers will come to a long-term equilibrium with the ceiling, walls and floor of the storage unit.

An investigator made post-discovery on-site air temperature and RH measurements inside the storage unit on January 9 through 12, 2003 using a handheld humidity/temperature monitor with internal sensors. The monitor's temperature accuracy was reported to be +/– 2 °F and RH accuracy was reported to be +/– 5% for values between 40% and 70% and +/–8% for RH below 40% or above 70%. Daily maximum and minimum air temperature and RH were recorded in the unit for four days. The difference between the average air temperature measurements made on-site and those at ASOS1 showed wide variability: 1 and 4 °F warmer on the first two days and 7 and 15 °F warmer on the last two days. Two possible explanations were explored to account for the wide variability in temperature differences:

1. *Temperature lag explanation*: Since the storage unit is enclosed, could the heat content of the storage unit and difference in location cause shifts in the timing of the maximum and minimum temperatures air temperatures in the unit compared to ASOS1? A cold front passed over the area on 9 January with temperatures dropping the entire period. After this front passed over, the air temperatures at the airport decreased faster than those in the unit as would be expected from an uninsulated building without windows or open doors (which would allow heat from the storage unit to be lost to the outside air). However, since cloud cover dominated the period with clearing only occurring on the 12th, neither heat storage lags or differences in solar heating between the ASOS1 station and the unit could not explain the wide range of conditions calculated.

2. *In-unit measurement problem*: Since the maximum and minimum air temperature and RH measurements made in the storage unit were manually recorded daily and the storage units is small, could the daily visitation by the investigator to read the logger contributed measurably to the amount of heat and moisture in the unit? During the winter, entry into the unit would cause heat to be lost when the door was open and heat and moisture added to the inside air volume due to the breathing of the investigator during the visit. We would expect the specific humidity of the air between the ASOS1 station and the urban center should be nearly the same. However, the humidity in the unit was greater than that at ASOS1 (excess humidity in the unit). To determine if the investigator could have added the excess humidity in the unit, we assumed that the maximum temperature and humidity for the day occurred at the time of the investigator's presence and calculated the visit duration needed to account for the higher humidity in the unit if solely due to the investigator's breath. The time intervals required to add breath humidity to the air was then compared against the excess heat in the unit relative to ASOS1. The elevated humidity and air temperature in the unit could be explained by a reasonable visit duration of the investigator. In addition, it appeared that the door to the unit was not always closed after entering: the investigator likely left the door open on the 9th and 10th but closed it on the 11th and 12th. While not conclusive, this effectively negated confidence in using the in-unit air temperature measurements in determining PMI.

Weather conditions over the period 6 through 17 October 2002 showed clear skies on one day, partly cloudy skies on three days, and cloudy skies on 8 days. Cloud cover was largely associated with frontal activity. The air temperature within the storage unit was therefore assumed not to be influenced by solar radiation heating the roof or the SSW wall of the unit (the other walls were against other units or facing away from the sun). Since cloud cover was mostly present with weather associated with frontal systems, the cloud cover would be expected to be low and the radiative heat loss to the sky from the roof during the night and day small. The ADD was estimated from the unadjusted ASOS1 temperature.

Conclusion and potential impact of errors

Assuming the bodies were always in the garbage containers and always in similar environments to the storage unit, the ADD (base 50 °F) associated with PMI for the male and female between the 7th and 17th of October was 106 °F day. This was in contrast to the estimate made by an entomologist for the prosecution (177 °F day) who estimated the ADD by adding the average difference between the on-site and ASOS1 temperatures made in 2003 of 7 °F to all ASOS1 air temperature measurements. The 106 °F day (bounded by potential sensor errors by 90 °F day to 122 °F day) ADD estimate was adequate for to explain the reported insect development on both corpses.

This ADD estimate however had limited usefulness: First, the on-site measurements not only were possibly defective but also they were done in a difference season with different characteristic weather, and second since it is clear that the bodies were moved and it is not known where the bodies were stored prior to their being moved to the storage unit. Although the bodies were likely in the storage unit between 11 and 17 October, the time of death was purported to be 6 October. Lack of knowledge of where the bodies were before 11 October limited the usefulness of this assessment.

Case 7: Corpse discovered in an unheated apartment building in Spring

A dismembered and partially decomposed male was discovered under clothing in a plastic bag in the closet of unheated second floor apartment on 18 May 1988. The closet door was closed. The downstairs apartment smelled a foul odor from their closet directly under this closet on the 12th or 13th and observed a discoloring of the closet ceiling. The victim was last seen alive on 6 May, 1988. The convicted defendant reportedly had confessed to the murder but evidence at the time of the initial trial, as well as in a 2000 post-conviction proceeding, suggested that the homicide occurred between 7 and 9 May-when the convicted defendant was in police custody on a separate charge but was released from custody on 11 May 1988. A new trial was deemed warranted in post-conviction proceedings but instead prosecutors dismissed the charges and the freed inmate then sued the city where he was convicted. An issue to be addressed in the civil suit was the validity of the initial estimate of when homicide occurred.

Approach

Estimate the ADD inside a 2nd floor apartment closet of an apartment within an urban area in proximity to a large Lake. An assessment of the local influence of the lake and the surrounding urban area was needed to model the outside of the building. A model of the indoor building microclimate was then needed to estimate the ADD at the corpse. Post-discovery on-site measurements made 15 years after the homicide were used to aid in the development of a building energy budget model.

Sources of data and evaluation of quality

- Hourly ASOS weather records at ASOS1 (6670 ft. asl and 13 miles from the apartment), ASOS2 (605 ft asl. and 9.5 miles from the apartment), and AWOS3 at the lake shore (603 ft asl.) from 6 am to 9 pm LT.
- Water temperature measurements every two hours at the city's drinking water intake two miles into the nearby lake with an assumed accuracy of +/− 2 °F.
- Hourly SAMSON solar radiation estimates with a reported error of +/− 4%
- Hourly post-discovery on-site air temperature measurements made in the apartment and in the closet in 2003. The sensor used was unknown.

Outside-building air temperatures: The apartment was within a densely populated urban area and 3 miles from the city center (population of about 3 million) in an area of high-density housing of less than four stories with an elevation of 604 ft. asl. The surrounding 22 square miles had urban land use and essentially flat. The apartment was 2.6 mi. from the shoreline while ASOS1 and ASOS2 were 12.9 and 8.2 mi from the shoreline (respectively). The ASOS2 station was within the urban area while the ASOS1 station was at the edge of the urban area.

The relative influence of the urban surroundings and the nearby lake were assessed. The water temperatures of the lake were less than the maximum air temperatures and greater than the minimum temperatures at both ASOS1 and ASOS2 airports for the period of 1 May 1988 through 18 May 1988. Lake water temperatures increased from 51 °F to 54 °F between 8 and 18 May 1988 were lower than the daily maximum air temperature and higher than the daily minimum air temperature at both locations. An expert witness for the defense estimated a rate of change in air temperature from the coast during days with distinct lake breeze of 0.4 °F to 2.3 °F per mile over the period of 6 May through 18 May 1988. Given the distance from the shore, the outside air temperature at the apartment would be 3 °F cooler due to the local breeze. However, the daily average air temperatures between 8 and 18 May 1988 averaged 3°F higher at ASOS2, nearer the lake, than ASOS1 regardless of an apparent lake breeze. These higher temperatures indicate the local climate of the urban area dominated the outdoor temperatures at the apartment building and negated the need to consider the lake breeze.

Within-building air temperatures: The 800 square foot apartment was modeled as four zones: front stairway outside apartment door, apartment excluding the closet in which the body was found, the closet where the body was found, and the back porch. A fifth zone was defined as the outside air. The hourly apartment microclimate model included:

- The solar radiation heat load of the outer walls of the apartment and the penetration of the solar radiation through the windows.
- Heat storage by the brick walls and heat exchanges and conduction of heat through the walls. The floor and ceiling of the second-floor apartment was assumed to be in equilibrium with the adjoining apartments.
- Convective heat exchange between the four zones in the apartment.
- Infiltration of the outside air into the apartment through widows. The influence of the surrounding trees and buildings on the infiltration was estimated given the 2003 on-site in-apartment air temperatures and the winds and air temperature at ASOS1.

The model was developed and model error estimated using the 2003 on-site measurements (Figure A7.1). The model underestimated the hourly apartment air temperature by 0.9 °F and the hourly closet air temperature by 0.3 °F (Figure A7.1). The overall model error was estimated at +/− 2 °F after including the error in outside temperature estimation.

The model was then applied to the 1988 outside air temperature estimates (Figure A7.2). The estimated average daily temperature in the apartment between 8 and 18 May 1988 was estimated to be 78.4 °F; 14.9 °F higher than

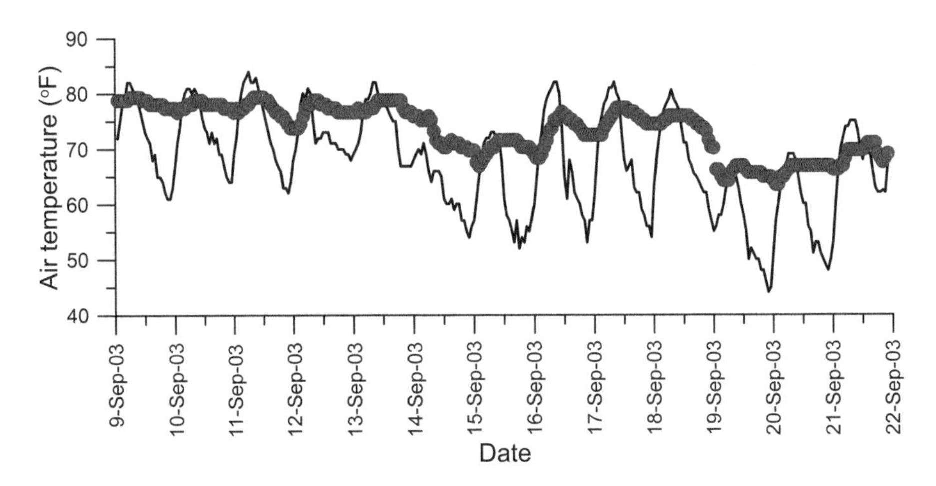

Figure A7.1 Post-discovery time series of estimated temperatures outside (black line) and measured on-site temperatures inside the closet where the body was found (red circle). Windows in apartment were open.

Data sources: NOAA National Climate Data Center, post-discovery data made by local law enforcement.

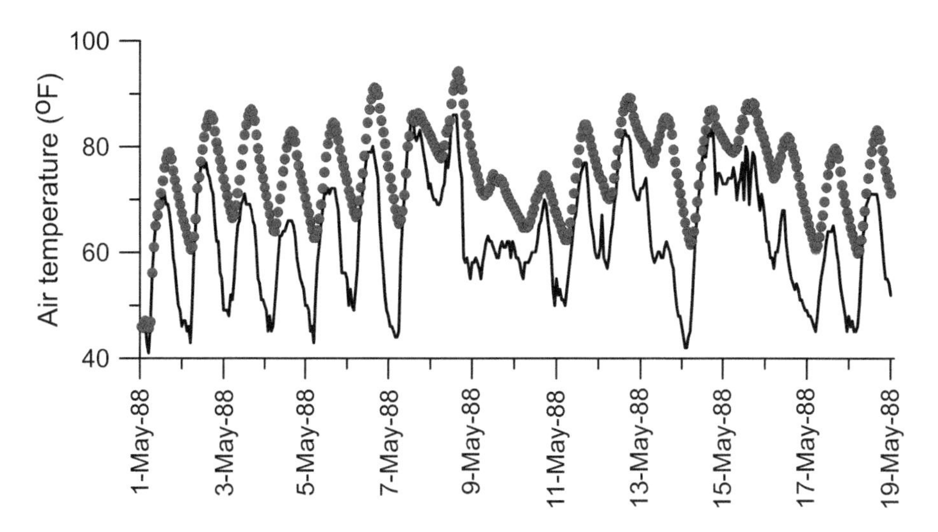

Figure A7.2 Hourly time series of estimated air temperatures outside (black line) and inside the closet where the body was found (solid red circle). Windows in apartment were closed.

Data source: NOAA National Climate Data Center

the air temperature measurements made at ASOS1. The average modeled daily temperature in the closet where the body was located was 0.4°F higher than the air in the apartment.

The large differences in the daily range in closet air temperatures (Figures A7.1 and A7.2) was due to the difference in infiltration between the period of interest and post-discovery measurements. The windows were painted shut in 1988 (based on the recollection of the investigators) while they were open during the 2003 on-site measurements.

Conclusions and impact

Between 18 May and 6 May 1988, the ADD (ADD base 50 °F) for the corpse in the closet was estimated at 380 °F d while that for the outside air, used in the initial PMI estimate in the first trial, was 181 °F d. The development was used to determine the likely time of death. The forensic entomologist indicated that at least 112 °F d were needed for the development of the *Phaenicia sericata* larvae observed on the body. Consequently, the microclimate modeling shifted the dates of likely death from the between the 8th and 11th of May compared to that when estimated directly from the ASOS1 air temperatures of later than 14th or 15th of May. This shift in dates of likely murder corresponded with a change in dates from when the freed male was in custody on a separate charge to after his release and explained the noticeable odor in the closet and discoloring of the closet ceiling of the apartment below on the 12th or 13th of May.

Case 8: Corpse remains discovered at two times (fall and following fall) in mountainous terrain

The remains of a male corpse were found at two different times in two different locations in the Rocky Mountains of the USA. The first discovery on 5 October 2011, by US Government employees, consisted of torn-open black plastic bags containing a partial corpse and some bones scattered on the ground in a forest and below an embankment near a closed camping area at 5308 ft asl (Location 1). A second group of torn-open plastic bags with bones and a skull were found by a man walking his dog on 25 September 2012 in thick tall grass in a wooded area at 5167 ft. asl (Location 2). Analysis revealed that these were from the same corpse. The deceased was last seen 26 June 2011.

Approach

Evaluate the ADD and average daily humidity for the period of 26 June 2011 to 5 October 2011 for the remains at Location 1 (time period 1) and for the period of 26 June 2011 to 25 September 2012 for the remains at Location 2 (time period 2). The modeled microclimates need to consider the influences of mountainous terrain, a snow-covered landscape, and local influences of sloped terrain, woodland, and forest.

Sources of data

- Records of 'Summary of the Day' measurements at COOP1 station (4851 ft. asl, 4 mi distant from Location 1 and 8.2 mi distant from Location 2).
- Hourly AWOS measurement records at AWOS1 (3865 ft. asl, 13.4 mi distant from Location #1 and 19.2 mi distant from Location 2).
- Hourly measurement records at a State Dept. of Transportation Road Weather Information Service (RWIS) records (6320 ft. asl, 2.5 mi distant from Location 1 and 3.8 mi distant from Location 2).
- National Weather Service National Operational Hydrological Remote Sensing Center (NOHRSC) Regional Snow analysis.

Assessment of measurement and modeling errors

In mountainous terrain, the most representative measurements may be those closest in elevation or closest in distance. While the closest station to Locations 1 and 2 was the RWIS station, there was no documentation available to know the accuracy or maintenance history of the sensors and consequently the climate record could not be used. The closest station in elevation to Locations #1 and 2 was COOP1; 457 ft below Location 1 and 316 ft below Location 2.

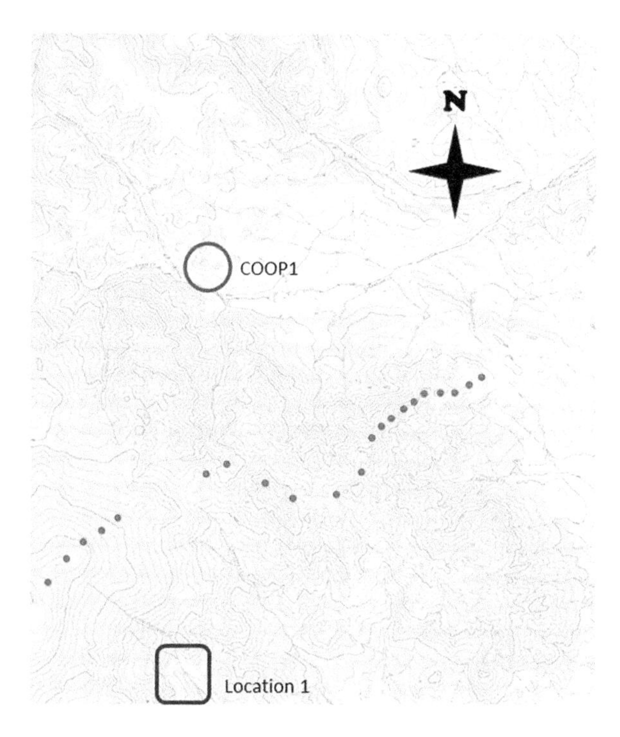

Figure A8.1 **Terrain between COOP1 and Location 1**. COOP1 (circle) and Location 1 (square) are separated by a ridge (dotted green line) giving both locations the nominal westerly airflow across the mountains. Elevation lines at 80 ft intervals.

Map source: US Geological Survey, https://www.usgs.gov/programs/national-geospatial-program/topobuilder

Although there is a ridge between COOP1 and Location 1 (Figure A8.1), the ridge would not influence the nominally westerly flow of air reaching COOP1 and Location 1 differently. Location 2 is upwind of COOP1 but also located in a relatively level area (Figure A8.2). The AWOS1 station was 1443 ft below Location 1 and 19.3 mi from and 1302 ft below Location 2.

The daily temperature measurements at COOP1 station were determined to be the most representative of the conditions at both Location 1 and 2. Maximum and minimum air temperature observations at COOP1 were only partially present or entirely missing on 4 days during time period 1 and was missing for 256 days over time period 2. In addition, the daily temperatures at COOP1 were invalidated on 12 days over the period of record for Location 2 due to inconsistencies. The maximum and minimum air temperature and dew point temperature record was complete for both time periods at the AWOS1 station. Consequently, missing daily average temperature estimates at COOP1 were estimated from daily maximum and minimum temperature measurements made at AWOS1 after adjustments for elevation, clouds, snowpack, and local microclimates. The estimated local climate at Locations 1 and

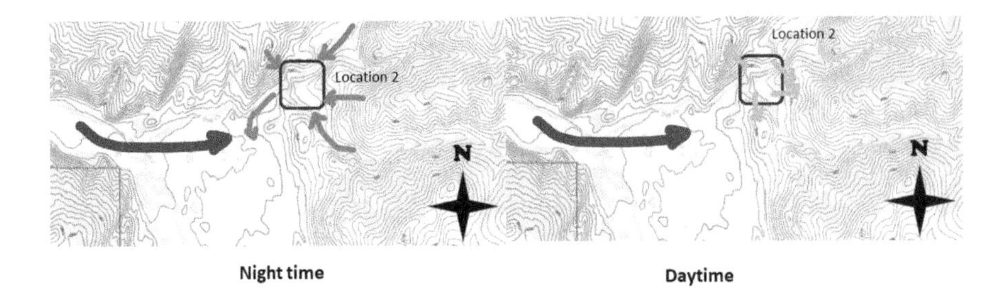

Figure A8.2 Topography around the Location 2. Note relatively flat terrain around discovery area Location 2 (box) and the high hills to west. The prevailing winds as well as the expected surface valley/slope winds during night time (blue arrows) and daytime (yellow arrows) are indicated. Elevation lines at 80 ft intervals.

Map source: US Geological Survey, https://www.usgs.gov/programs/national-geospatial-program/topobuilder

2 were then estimated from the daily average COOP1 temperature records and the daily average humidity measurements of AWOS1 after correcting for elevation, clouds, snowpack, and local microclimates.

Adjustments to climate record

Altitude correction: Unmeasured daily average temperatures at COOP1 were estimated from the average of the daily minimum and maximum air temperatures at AWOS1 after adjusting for differences in elevation. The error in estimating the missing COOP1 daily average temperatures from measurements at AWOS1 was determined from all daily paired measurements between 1 June 2011 and 30 September 2012. The estimated daily average temperatures at COOP1 were on average 1.6 °F lower than the altitude-corrected AWOS1 temperatures. Estimates of COOP1 from AWOS1 daily average temperature were adjusted for this bias. The error in the bias-corrected estimates of the daily average temperature at COOP1 based on AWOS1 was +/− 5.5 °F.

The estimated error in daily average RH from AWOS1 measurements was estimated assuming the bias- adjusted altitude corrected AWOS1 temperatures had an error of +/− 5.5 °F and a RH error of +/− 7%.

Cloud correction: Cloud cover at COOP1 was assumed identical to that measured at AWOS1 13 mi to the east. Although the air cools at a slower rate with increasing altitude within clouds than in the clear sky, the clouds did not significantly change the bias in average daily temperature estimation for COOP1 based on AWOS1. Clouds were estimated to be at the ground surface at Locations #1 and #2 43% of the time. It was assumed that fog (with a RH of 100%) had formed at COOP1 if the altitude-corrected temperature was less than the dew point temperature at AWOS1 but the AWOS1 cloud height measurement did not indicate a cloud should be present at the ground at the COOP1 elevation. If a cloud was indicated to be at ground level at COOP1 elevation, the daily average RH was assumed to be 99% if daily average cloud cover was greater than 50% and 94% when daily average cloud cover was less than 50%. The error of the daily average RH based on cloud presence was estimated at 5%.

Snowpack correction: Assuming the remains were left at Location 2 at the same time that they were left at Location 1, the remains overwintered at Location 2. Snow cover was estimated from imagery provided at the NOHRSC 'Regional Snow analysis' website. Snow cover at Location 2 was assumed if the estimated depth was at least 0.4 inches on a given day. Snow cover at Location 2 was less frequent than expected for the mountainous region in general due to a higher peaks to the west of Location 2, removing the moisture from snow clouds rising over the peaks as the air blew in from the west. The RH was assumed to be 100% when a snowpack was estimated to be present at Location 2.

Local microclimate correction: It was assumed that the bags were torn open shortly after being deposited. Remains found at Location 1 were over an embankment from the paved road and in a heavily wooded area. Location 1 was situated on a gradual slope (Figure A8.3). Some slope valley winds may influence the air temperatures during the night, especially down valley flow (Figure A8.3). The heating of the forest canopy likely limits daytime slope flow at the ground. Location 1 was estimated to have an average temperature 1° F cooler than COOP1.

Remains found at Location 2 were in a shallow lightly-wooded valley (Figure A8.2). Some slope valley winds may influence the air temperatures during the night (Figure A8.2). The open woods at this location would permit sunlight to the woods floor, although the remains were unlikely to be influenced by sunlight as they were partially buried in leaves and other dead vegetation. While soil temperatures likely influenced the daily average temperature of the partially-buried remains, no daily average soil temperature could be calculated without post-discovery measurements. It was assumed that the average temperatures in the understory litter and exposed bones were likely less than a degree different than the open above the litter. No air temperature corrections were made for the local environment at Location 2.

The average humidity within the understory litter was likely greater than that of the air when the litter and ground is moist and approximately that of the air when the litter and ground is dry. Since no measurements of litter or

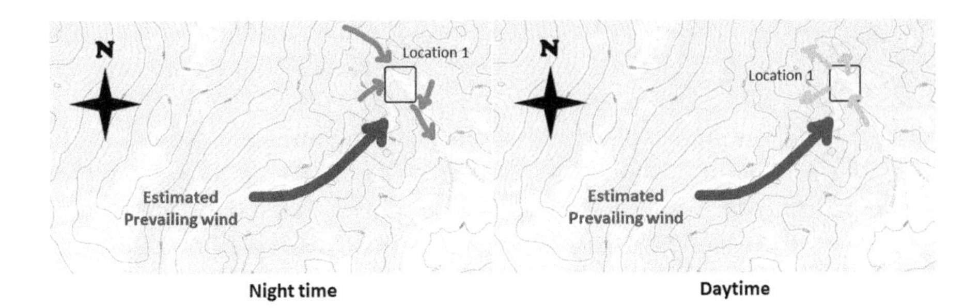

Figure A8.3 Topography west of the Location 1. Note high peaks to the west and a gradual slope towards creek the moving to the southeast. The region around Location 1 (indicated by box) was lower than that to the west. The prevailing winds as well as the expected surface valley/slope winds during night time (blue arrows) and daytime (yellow arrows) are indicated. Elevation lines at 80 ft intervals.

Map source: US Geological Survey, https://www.usgs.gov/programs/national-geospatial-program/topobuilder

ground wetness were available, the RH in the litter was assumed the same as that of the overlying air. Consequently, no microclimate adjustments were made to the altitude-corrected humidity for the remains at either Location 1 or 2.

Summary of method: Daily average air temperatures were estimated from COOP1 when maximum and minimum air temperature measurements were available and based on altitude-corrected daily average temperature measured at AWOS1 adjusted by the average difference between the estimated and measured daily average temperatures at COOP1 when both temperatures were available. Daily average temperatures at the two Locations were estimated from the COOP1 daily average temperature after altitude-correction. A local terrain temperature adjustment was made for Location 1 and a snowpack temperature adjustment was made for Location 2. Daily average RH at both Locations was estimated by assuming the air at the Locations had the same daily average water content of the air at AWOS1 unless:

a. the air temperature at a Location was less than the daily average dew point temperature measured at AWOS1 and RH was assumed to be 100%,
b. the bottom of the clouds measured at AWOS1 was lower than the Location altitude and RH depended on the daily average cloud cover, or
c. Location #2 appeared to be snow-covered and RH was assumed to be 100%.

Conclusion and potential impact of errors

The average air temperature and RH at Location 1 from time of last known sighting to remains recovery (252 days) was 60 °F +/– 0.4 °F and 43% +/– 1% respectively. The average air temperature and RH at Location 2 from time of last known sighting to remains recovery (638 days) was 47°F +/– 0.2 °F and 48% +/– 1% respectively. The long periods between last sighting and remains discovery at each location resulted in a relatively low ADD estimate accuracy. The ADD (base 50 °F) was 1003 °F d with possible range of 494 °F d to 1594 °F d at Location 1 and 1355 °F d with possible range of 704 °F d to 2115 °F d after the 2011–2012 winter thaw resulted in air temperatures above 50°F at location 2.

Index

Pages in *italics* refer to figures and pages in **bold** refer to tables.

R